MAKING MATH LEARNING FUN FOR INNER CITY SCHOOL STUDENTS

Edited by
Glendolyn Duhon-Jeanlouis
Alice Duhon-McCallum
Ashraf Esmail

Issues in Black Education Series

University Press of America,® Inc.
Lanham · Boulder · New York · Toronto · Plymouth, UK

Copyright © 2011 by
University Press of America,® Inc.
4501 Forbes Boulevard
Suite 200
Lanham, Maryland 20706
UPA Acquisitions Department (301) 459-3366

Estover Road
Plymouth PL6 7PY
United Kingdom

All rights reserved
Printed in the United States of America
British Library Cataloging in Publication Information Available

Library of Congress Control Number: 2010933283
ISBN: 978-0-7618-5317-6 (paperback : alk. paper)
eISBN: 978-0-7618-5318-3

∞™ The paper used in this publication meets the minimum
requirements of American National Standard for Information
Sciences—Permanence of Paper for Printed Library Materials,
ANSI Z39.48-1992

We are dedicating this book to Dr. Glendolyn Duhon Jeanlouis sons, Charles and Jordan, who have always seen in her what she may not have seen in herself.

Contents

Series Foreword by Dr. Abul Pitre............................. ix
    References.. xiv
Foreword by Dr. Press Robinson............................. xvii
Preface by Dr. Glen Duhon-Jeanlouis......................... xix
Acknowledgments.. xxi
Chapter One – Making Mathematics Fun in the Classroom.......... 1
                  Glendolyn Duhon-Jeanlouis
    Multiplication... 3
    Problem Solving Strategies................................ 5
    Different Ways of Checking for Understanding.............. 6
    Personalized Word Problems................................ 7
    Charts and Graphs... 7
        Line Graphs: Charting the Weather.................... 7
        Grade Graphing....................................... 8
        Bar Graphs... 8
        M&M® Graphs.. 8
        Class Test Scores Graphs............................. 8
        Circle Graphs.. 9
    Measurement... 9
    Fractions.. 10
        Reducing Fractions.................................. 10
        Fraction Matches.................................... 11
        Adding and Subtracting Like Fractions............... 11
        Adding and Subtracting Unlike Fractions............. 11
    Conclusion... 12
    References... 12

Chapter Two – Preparing Students for Learning:
    Considering the Emotions............................... 13
        Melba Venison, Rose Duhon-Sells, and Mattie Spears
    References.. 21
Chapter Three – Factors Influencing Student Attitudes, Behaviors,
    and Learning Outcomes in the Statistics Course............. 23
        Lisa Eargle, Ashraf Esmail, and Jas Sullivan
    Introduction... 23
    Literature Review...................................... 24
    Data and Methods...................................... 26
    Findings.. 26
    Factors Influencing Student Learning Outcomes............. 27
    Instructional Procedures................................ 27
        Setting the Stage for the Course...................... 27
        Course Lecture Procedures.......................... 29
    Instructor's Disposition, Attitude, and Behavior............ 31
    Classroom Physical Environment........................ 32
    Additional Factors..................................... 33
        Student Characteristics............................. 33
        Instructor Characteristics........................... 33
        Other Course Features............................. 33
    Conclusion.. 34
    References.. 34
Chapter Four – Multicultural Education Math Strategies for
    Today's Diverse Classrooms.............................. 37
        Alice Duhon-McCallum, Doris L. Terrell, and Gwendolyn Duhon
    References.. 42
Chapter Five – After-Work Hours Challenges of American Parents
    to Assist Their Children with Math Homework and Other
    Subject Matters.. 45
        Donna Grant
    Do Any of These Situations Sound Familiar................ 46
    How Parents Can Help With Homework.................. 47
    Why Is It Important For My Child to Learn Math.......... 48
    How Will My Child Learn Math?......................... 48

What Tips Can I (Parents) Do To Help My Child With Math?. . 49
    Make Math Part of Your Child's Day. 49
        Encourage Your Child To Give Explanations. 50
Teaching Math to Young Children. 50
Foundations. 53
Why Is It Important For My Child To Learn Math. 59
Knowing How To Do Math Makes Our Day-to-Day
    Lives Easier!. 59
    How Will My Child Learn Math?. 59
What Tips Can I Use To Help My Child?. 60
    Be Positive About Math!. 60
    Make Math Part Of Your Child's Day. 60
    Encourage Your Child To Give Explanations. 61
What Math Activities Can I Do With My Child?. 61
    1. Understanding Numbers. 61
    2. Understanding Measurements. 62
    3. Understanding Geometry. 63
    4. Understanding Patterns. 64
    5. Understanding and Managing Data. 65
Where Can I Get Help. 66
    Your Child's Teacher. 66
    Others Who Can Help. 66
    References. 66
Index 67
Editors. 69
Contributors. 71

SERIES FOREWORD

Historically, the state of Black education has been at the center of American life. When the first Blacks arrived to the Americas to be made slaves, a process of *mis-education* was systematized into the very fabric of American life. Newly arrived Blacks were dehumanized and forced through a process that has been described by a conspicuous slave owner named Willie Lynch as a "breaking process": "Hence the horse and the nigger must be broken; that is, break them from one form of mental life to another – keep the body and take the mind" (Hassan-EL, 2007, p. 14). This horrendous process of breaking the African from one form of mental life into another included an elaborate educational system that was designed to kill the creative Black mind. Elijah Muhammad called this a process that made Black people blind, deaf, and dumb – meaning the minds of Black people were taken from them. He proclaimed, "Back when our fathers were brought here and put into slavery 400 years ago, 300 [of] which they served as servitude slaves, they taught our people everything against themselves" (Pitre, 2008, p.6). Woodson (2008) similarly decried,

> Even schools for Negroes, then, are places where they must be convinced of their inferiority. The thought of inferiority of the Negro is drilled into him in almost every class he enters and almost in every book he studies (p. 2).

Today, Black education seems to be at a crossroads. With the passing of the *No Child Left Behind Act of 2001,* schools that serve a large majority of Black children have been under the scrutiny of politicians who vigilantly proclaim the need to improve schools while not realizing that these schools were never intended to educate or educe the divine powers within Black people. Watkins (2001) posits that after the Civil War, schools for Black people – particularly those in the South – were designed by wealthy philanthropists. These philanthropists designed "seventy five years of education for blacks" (pp. 41-42). Seventy-five years from 1865 brings us to 1940, and today we are sixty-nine years removed from 1940. The sum of these numbers does not equal seventy-five years of scripted education; to truly understand the plight of Black education, one has to consider the historical impact of seventy-five years of scripted education and its influence on the present state of Black education.

Presently, schools are still controlled by ruling class Whites who hold major power. Woodson (2008) saw this as a problem in his day and argued, "The education of the Negroes, then, the most important thing in the uplift of Negroes, is almost entirely in the hands of those who have enslaved them and now segregate them" (p.22). Here, Woodson cogently argues for historical understanding: To point out merely the defects as they appear today will be of little benefit to the present and future generations. These things must be viewed in their historic setting. The conditions of today have been determined by what has taken place in the past...(p.9). Watkins (2001) summarizes that the "white architects of black education...carefully selected and sponsored knowledge, which contributed to obedience, subservience, and political docility" (p. 40). Historical knowledge is essential to understanding the plight of Black education.

A major historical point in Black education was the famous *Brown v. the Board of Education Topeka Kansas*, in which the Supreme Court ruled that segregation deprived Blacks of educational equality. Thus, schools were ordered to integrate with all deliberate speed. This historic ruling has continued to impact the education of Black children in myriad and complex ways.

To date, the landmark case of *Brown v. the Board of Education Topeka Kansas* has not lived up to the paper that it was printed on. Schools are more segregated today than they were at the time of the *Brown* decision. Even more disheartening is that schools that are supposedly desegregated may have tracking programs such as "gifted and talented" that attract White students and give schools the appearance of being integrated while actually creating segregation within the school. Spring (2006) calls this "second-generation segregation" and asserts: Unlike segregation that existed by state laws in the South before the 1954 *Brown* decision, second generation forms of segregation can occur in schools with balanced racial populations; for instance, all White students may be placed in one academic track and all African American or Hispanic students in another track (p. 82). In this type of setting, White supremacy may become rooted in the subconscious minds of both Black and White students. Nieto and Bode (2008) highlight the internalized damage that tracking may have on students when they say students "may begin to believe that their placement in these groups is natural and a true reflection of whether they are 'smart,' 'average,' or 'dumb'" (p. 119). According to Oakes and Lipton (2007),

> African American and Latino students are assigned to low-track classes more often than White (and Asian) students, leading to two separate schools in one building – one [W]hite and one minority (p. 308).

Nieto and Bode (2008) argue the teaching strategy in segregated settings "leaves its mark on pedagogy as well. Students in the lowest levels are most likely to be subjected to rote memorization and static teaching methods" (p. 119). These findings are consistent with Lipman's (1998):

> scholars have argued that desegregation policy has been framed by what is in the interest of [W]hites, has abstracted from excellence in education, and has been constructed as racial integration, thus avoiding the central problem of institutional racism (p. 11).

Hammond (2005) is not alone, then, in observing that "the school experiences of African American and other minority students in the United States continue to be substantially separate and unequal" (p. 202).

Clearly, the education of Black students must be addressed with a sense of urgency like never before. Lipman (1998) alludes to the crisis of Black education, noting that "The overwhelming failure of schools to develop the talents and potentials of students of color is a national crisis" (p.2). In just about every negative category in education, Black children are over-represented. Again Lipman (1998) alludes, "The character and depth of the crisis are only dimly depicted by low achievement scores and high rates of school failure and dropping out" (p. 2). Under the guise of raising student achievement, the *No Child Left Behind Act* has instead contributed to the demise of educational equality for Black students. Hammond (2004) cites the negative impact of the law:

> The Harvard Civil Rights Project, along with other advocacy groups, has warned that the law threatens to increase the growing dropout rate and pushout rates for students of color, ultimately reducing access to education for these students rather than enhancing it (p. 4).

Asante (2005) summarizes the situation thus: "I cannot honestly say that I have ever found a school in the United States run by whites that adequately prepares black children to enter the world as sane human beings...an exploitative, capitalist system that enshrines plantation owners as saints and national heroes cannot possibly create sane black children" (p. 65). The education of Black students and its surrounding issues indeed makes for a national crisis that must be put at the forefront of the African American agenda for liberation.

In this series, *Issues in Black Education,* I call upon a wide range of scholars, educators, and activists to speak to the issues of educating Black students. The series is designed to not only highlight issues that may negatively impact the education of Black students but also to provide possibilities for improving the quality of education for Black students. Another major goal of the series is to help pre-service teachers, practicing teachers, administrators, school board members, and those concerned with the plight of Black education by providing a wide range of scholarly research that is thought-provoking and stimulating. The series will cover every imaginable aspect of Black education from K-12 schools to higher education. It is hoped that this series will generate deep reflection and catalyze action-praxis to uproot the social injustices that exist in schools serving large numbers of Black students.

In the past, significant scholarly research has been conducted on the education of Black students; however, there does not seem to be a coherent theoretical approach to addressing Black education that is outside of European dominance. Thus, the series will serve as a foundation for such an approach – an examination of Black leaders, scholars, activists, and their exegeses and challenge of power relations in Black education. The idea is based on the educational philosophies of Elijah Muhammad, Carter G. Woodson, and others whose leadership and ideas could transform schools for Black students. One can only imagine how schools would look if Elijah Muhammad, Carter G. Woodson, Marcus Garvey, or other significant Black leaders were in charge. Additionally, the election of Barack Hussein Obama as the first Black president of the United States of America offers us a compelling examination of transformative leadership that could be inculcated into America's schools. The newly elected president's history of working for social justice, his campaign theme of "Change We Can Believe In," and his inaugural address that challenged America to embrace a new era are similar to the ideas embodied in *Critical Black Pedagogy in Education.*

This series is a call to develop an entirely new educational system. This new system must envision how Black leaders would transform schools within the context of our society's diversity. With this in mind, we are looking not only at historical Black leaders but also at contemporary extensions of these great leaders. Karen Johnson (in press) describes the necessity for this perspective:

> There is a need for researchers, educators, policy makers, etc. to comprehend the emancipatory teaching practices that African American teachers employed that in turn contributed to academic success of Black students as well as offered a vision for a more just society.

Freire (2000) also lays a foundation for critical Black pedagogy in education by declaring, "it would be a contradiction in terms if the oppressors not only defended but actually implemented a liberating education" (p. 54). Thus, critical Black pedagogy in education is a historical and contemporary examination of Black leaders (scholars, ministers, educators, politicians, etc.) who challenged the European dominance of Black education and suggested ideas for the education of Black people.

One of the major disciplinary areas of concern in the education of Black students is mathematics. Across the country educators are faced with trying to raise the math scores of Black students. Even more striking is that the current high stakes testing era which is heavily focused on math has become more complex meaning that unlike math tests in years past that required students to do math equations alone the current testing in math uses word problems that entail the complex use of language. Additionally, more complex levels of mathematic concepts are being introduced to children in the early grades. Educators can no longer focus on reading scores as math has become the major gatekeeper that is negatively impacting Black students. In too many cases educators and students have become fearful of math. Clearly educators and students alike have forgotten that it was great Black mathematicians that gave the world some of the most innovative mathematical concepts. In fact, some have noted that there are 59,999 additional books of math that the original Black scientists have kept hidden from the world at large. While teaching math is an endemic problem confronting the whole of America it is illuminated in inner cities where educators and students are faced with a plethora of issues the make learning math problematic. One factor is that the teaching of mathematics is disconnected from the cultural experiences of too many Black students. In an effort, to provide more culturally responsive teaching strategies the contributors in this timely book, *Making Math Learning Fun For Inner City School Students* have provided creative and compelling ways to engage inner city students in the study of math. While reading the numerous ideas presented in the book educators will be amazed at the wide range of activities that can be easily transferred to the classroom or the kitchen table. The strategies and concepts in the book provide essential elements for unlocking the creative mind in students who may be struggling with math. The book is indeed compelling and offers new approaches to teaching math that even parents, tutors, and community programs can use.

Making Math Learning Fun For Inner City School Students is a welcome addition to the literature on Black education. Similar to Joyce King's (2005) *Black Education: A Transformative Research and Action Agenda for the New Century,* this book addresses research issues raised in *The Commission on Research in Black Education* (CORIBE). Like CORIBE's agenda, *Making Math Learning Fun For Inner City School Students* focuses on "using culture as an asset in the design of learning environments that are applicable to students' lives and that lead students toward more analytical and critical learning" (p. 353). The book is indeed provocative, compelling, and rich with information that will propel those concerned with equity, justice, and equality of education into a renewed activism.

Abul Pitre
Series Editor

REFERENCES

Asante, K. (2005). *Race, rhetoric, & identity: The architecton of soul.* Amherst, NY: Humanity Books.

Freire, P. (2000). *Pedagogy of the oppressed.* New York: Continuum.

Hammond-Darling, L. (2004). From "separate but equal" to "no child left behind": The collision of new standards and old inequalities. In Meier, D. & Wood, G. (Eds.). *Many children left behind: How the no child left behind act is damaging our children and our schools (pp. 3-32).* Boston: Beacon Press.

Hammond-Darling, L. (2005). New standards and old inequalities: School reform and the education of African American students. In King, J. (Ed.). *Black education: A transformative research and action agenda for the new century (pp. 197-224).* Mahwah, NJ: Lawrence Erlbaum Associates.

Hassan-EL, K. (2007). *The Willie Lynch letter and the making of slaves.* Besenville, IL: Lushena Books.

Johnson, K., & Pitre, A. (Eds.). (in press). African American women educators: *A critical examination of their pedagogies, educational ideas, and activism from the nineteenth to the mid-twentieth centuries.* Lanham, MD: University Press of America.

King, J.E. (Ed). (2005). *Black education: A transformative research and action agenda for the new century.* Mahwah, NJ: Lawrence Erlbaum Associates.

Lipman, P. (1998). *Race and the restructuring of school.* Albany, NY: SUNY Press.

Nieto, S., & Bode, P. (2008). *Affirming diversity: The sociopolitical context of multicultural education* (5th ed.). Boston: Allyn and Bacon.

Oakes, J., & Lipton, M. (2007). *Teaching to change the world. (3rd Ed.).* Boston: McGraw-Hill.

Pitre, A. (2008). *The education philosophy of Elijah Muhammad: Education for a new world.* (2nd ed.). Lanham, MD: University Press of America.

Spring, J. (2006). *American education.* New York: McGraw-Hill.

Watkins, W. (2001). *The White architects of Black education: Ideology and power in America 1865-1954.* New York: Teachers College Press.

Woodson, C.G. (2008). *The mis-education of the Negro.* Drewryville, VA: Kha Books.

FOREWORD

The educational system in this country, based on state, regional, and national test scores has not accomplished the goal of providing a strong mathematic program for inner city school students. Therefore, many inner city students' perception of math is often negative and viewed as a subject that is beyond their ability. Thus, they believe there is no way they will be successful in earning good grades in math. The fact is, inner city students are capable of learning math as well as any other subject. The key to any student learning math is effective classroom teachers who believe in their ability to learn and have the essential skills to nuture them , but also to challenge, motivate and encourage them to do their best. When the teacher has confidence in the ability of students to conquer the challenge of learning math in an environment that is supportive and nurturing, they will learn. The curriculum for preparing classroom teachers needs a new paradigm, one that will prepare classroom teachers to develop strategies and techniques to gain inner city students' trust and respect, understand their circumstances, and reach out to them in a variety of ways. The inner city school population in this country is often misunderstood by educators and others. Often the inner city school students are discussed in the media in negative terms and teachers use that information to develop their perception about the students thereby not believing they have the ability to experience success in mathematics. The ability is there, but teachers need to develop their desire to learn.

This manuscript was developed by a group of professional educators/ researchers and individuals concerned about the academic success of inner city students and the importance of a strong, fun-filled, intellectually stimulating and motivating mathematics curriculum. This group has taken the time to develop a book focused on making learning mathematics fun for inner city school students. This book may be used as an essential tool for parents, teachers, and professionals at all levels to simplify mathematic concepts at a level that the subject matter will be clarified, understood and enjoyed. The subject of mathematics is most important in elementary schools because it is the stepping stone for students to move into other levels of computation. Classroom teachers in the inner city schools must understand and feel comfortable with math in order to spend their time and energy preparing to teach the subject.

Pre-math concepts start with children as early as two or three years old as their parents assist them while counting household objects as they play number games. Many inner city school students do not have the privilege of participating in pre-math activities with their parents. Often, inner city parents have to work several jobs just to make ends meet and have a limited amount of time to interact with their young children with pre-math or any other early stage learning activities. This book can be very helpful because it provides a body of knowledge that enables the parent to enjoy learning the math games with their children.

The subject of mathematics has been labeled with numerous myths. These myths have caused many teachers in elementary and middle schools to develop a fear of the subject. And, it is this fear that leads them to avoid it or teach it poorly, specifically in inner city schools. Unfortunately, many children overhear conversations from their parents about how they had problems with math while in elementary school. As a result, school-aged students immediately use that information as a justifiable excuse for not understanding or enjoying mathematics.

This book is a cutting-edge approach addressing inner city school students' success in learning and enjoying mathematics in K-12 schools.

Dr. Press Robinson
President
Duplichain University

PREFACE

America's inner city schools must meet the many challenges they face in the teaching of mathematics. New challenges in the way we approach the teaching of mathematics in public schools are needed in order to motivate students in inner city schools. The inner city student presents a new student population, one that is not understood by many teachers who were trained ten years ago. Their behavior is often mistakenly viewed as bad or disrespectful and the students are only living what they have learned. In many of their homes and communities that behavior is humorous and reinforced by laughter or complemented by family members viewing it is acceptable. Teacher education training programs must include effective teaching strategies appropriate for new teachers to understand and relate to the inner city student. This is very important, before they will be capable of teaching them to learn and enjoy mathematic concepts.

This book is written for teachers who are prepared to teach mathematics with success for all students. The focus is on identifying and creating engaging teaching strategies in mathematics, which will meet the different learning styles of the inner city students.

Encouraging students to think differently about mathematics is an extremely difficult challenge. Because, many students think mathematics is difficult to learn. The teacher must use teaching strategies that will cultivate the students' areas of interest in mathematics and ensure them, that they are capable of experiencing academic success in mathematic.

Understanding the learning styles of students, teachers can encourage students to engage in mathematics through different kinds of purposeful activities hence, focusing on the student performance and success in the area of mathematics. The key to improving mathematics in the inner city schools, lies with teachers, parents and students working together to provide quality work in mathematics, that enables students to learn what they need in order to be successful in school.

Dr. Glen Duhon-Jeanlouis

Acknowledgments

We would like to first thank God, and all who have impacted and encouraged us to successfully complete this book. We would like to thank, Dr. Rose Marie Duhon-Sells, our "Wind beneath My Wings." Her constant, and persistent urging accompanied with a strong belief fostered us into fruition of this task.

We would like to thank Dr. Gwendolyn Duhon for her belief and confidence in us and Patrick Caffery Duhon, Jr. for giving us a silent shoulder to lean on whenever needed.

We would like to thank Dr. Mary Addison and Mrs. Carolyn Figaro for continuously caring and believing in us. These ladies are truly an inspiration.

We would finally like to thank Dr. Glendolyn Duhon-Jeanlouis's husband, Wendell, and her sons, Charles and Jordan. Wendell gave her the freedom to pursue her dreams, and advice when asked. Her sons reminded her that she must follow her dreams, in order for them to aspire to do so.

CHAPTER ONE
MAKING MATHEMATICS FUN IN THE CLASSROOM

GLENDOLYN DUHON-JEANLOUIS

The academic subject of Math has been demonized in schools from K-12 grade into higher education. In social gatherings, lounges in industry, and any place adults gather you will hear someone make the comment "I have always had a problem in Math. When I was in elementary/high school, I hated Math." Consequently, adults who make these comments subconsciously make the same comment in the presence of their children or other children in their community. Those comments create an acceptable excuse for school-aged children not to focus or concentrate on being successful or earning good grades in Math. School-aged children must believe they have the ability to achieve A's and B's in any subject area, if they are willing to follow directions, work hard, and focus on academic success.

According to C. Jeet (October, 2008). Math is a set of interrelated concepts. One concept serves as the prerequisite for many other concepts. Math at early years of elementary school forms the basis for K-12 mathematics. For any student, success in mathematics at elementary school-level is critical as it leads to future success in relation to middle and high school mathematics, university entrance, etc.

To enable students to focus on more difficult tasks such as problem solving, proficiency in basic math concepts is vital. Otherwise, students will focus more on the basic skills rather than the task at hand; as a result their concentration deviates from the learning objectives of the present task.

Teaching and learning mathematics are complex tasks. The effect on student learning of changing a single teaching practice may be difficult to discern because of the simultaneous effects of both the other teaching activities that surround it and the context in which the teaching takes place. Research findings indicate that certain teaching strategies and methods are worth careful consideration as teachers strive to improve their mathematics teaching practices. (applesfortheteacher.com).

There are many innovative strategies for teaching Math and making the concepts relevant to the everyday lives of children. Math must be taught in a meaningful fun-filled context. Committed excellent classroom teachers who are dedicated and willing to spend the time conducting research, learning how the students in their class approach new learning concepts are capable of helping children to develop the perception that Math, as any other subject area, will become simple for the students to learn. All students are capable of earning above average grades in Math, and gaining the self-confidence needed to conquer any academic challenge including Mathematics.

According to D. Iverson (October, 2009), children SHOULD learn their basic facts, be able to do mental calculations, and master long division and other basic paper-pencil algorithms. Mathematics is a field of study that builds on previously established facts. A child that does not know basic multiplication (and division) facts will have a hard time learning factoring, primes, fraction simplification and other fraction operations, distributive property, etc. Basic algorithms of arithmetic are a necessary basis for understanding the corresponding operations with polynomials in algebra. Mastering long division precedes understanding how fractions correspond to the repeating infinite (non-terminating) decimals, which then paves way to understanding irrational numbers and real numbers. It all connects together!

For this reason, it is probably very wise to restrict calculator use in the lower grades until a child knows his/her basic facts and can add, subtract, multiply, and divide even large numbers with pencil and paper. THIS, in my opinion, can build number sense – as do mental calculations.

Educators and parents must join forces to help school-aged children eradicate the fear and uncomfortable feelings about Mathematics as a subject too difficult for anyone to achieve high academic grades. The classroom teacher is the academic leader of the classroom and the parent is the champion of their children's learning. Therefore, the united effort over time will remove the myth and change children's perceptions about Math as an unachievable subject area. One very innovative strategy is to teach Mathematics from an interdisciplinary approach helping children to identify the common thread between Math, Social Studies, Geography, History, English, and the Arts. All of these subject areas have mathematical elements in their discourse such as when teaching Geography, teachers could help children to recognize the mathematical elements when having a discussion of distances between one city and the other, sizes of the various populations in a given region, and size of one city compared to the other. Social Studies is an excellent subject that creates a glue among other subject areas, particularly Math when you address social issues such as the difference between a company employing ten people as opposed to one employing 100 people. In History, mathematical concepts are constantly measuring time, distance, historical events, icons and legends of the time. These are learning activities and discussions that will enable children to sit at the edge of their seats with curiosity and interest in learning more and begging to learn more as they identify the mathematical elements throughout the discussion.

One outstanding middle school teacher in the Texas public school system has proven that innovative mathematical concepts do make a difference in the educational experience of middle school students. The following Math activities will help the reader to recognize the necessity for creating innovative instruction for school aged-children from K-12 grade, but most important middle school. The National Department of Education conducted a survey that predicted children, who are in middle school and the learning foundation is not strong enough to endure learning new concepts such as Mathematics, were most likely to fail or drop out of school before graduating from the twelfth grade.

There are many strategies that the author uses in order to effectively teach Math. The author begins with constantly telling the students that they are capable of solving any problem that they put their mind to and that understanding Math is 50 percent attitude, and 50 percent practice.

MULTIPLICATION

The author believes that students should memorize their Math facts, due to the fact that they must know their times tables in order to solve the majority of Math problems.

At the beginning of each school year, students copy their multiplication facts from 2-12 two days per week for the first month of school. On Friday of each week, a fact drill test is administered. Students are given strips of sheets numbered from 1-20. The teacher calls out the math facts and the students are to write the correct answer for each corresponding number.

A related game is called multiplication baseball. The teacher assigns four areas designated for first, second, third, and home plate respectively. The class is divided equally into two teams. It is recommended that remaining students may act as score keepers of fact callers. Students are given approximately seven seconds to answer. The teacher may act as time-keeper or designate a student. The first team has Math facts called to them. If the students answer the correct product, that student then goes to the assigned base. When a student answers with an incorrect product, that team is given a strike. Once the team has three strikes, that team loses their turn, and the other team has Math facts assigned to them. After both teams have attained three strikes, the team with the highest score wins. If both teams do not earn any points, play resumes until the first team receives a point. Teams should be divided evenly based on skill level, as well as boy/girl ratio. This activity is well received by the students because they enjoy the competitive nature, the moving aspect, and the fact that they are not writing. It is also a way of getting students to know each other.

A second activity is Fact Toss. A large, soft ball or a bean bag is used. The students practice calling out their facts and line up in two teams facing each other horizontally. Students call out "3" if the students are practicing their multiples of 3. Then he/she throws the ball to the student across from him. The students are moving, and have to pay attention so that they will call out the correct fact. Normally, the game is started with multiples of 0 and end with multiples of 12. For those students who do not know a particular multiple, the other students tell them, or the group may repeat the multiples for that particular number.

"Fact Face-off" – The students are lined up in two vertical teams. Students are shown a fact card and have seven seconds to answer. If that student either does not answer or answers incorrectly, that student is out of the game and returns to his/her seat. The last student standing is the winner. The teams can be divided into girls against boys, or class periods against each other.

Students are reminded that multiplication is simply a shorter way of adding two or more numbers. It is strongly believed that accurate knowledge of multiplication facts of the foundation for all areas of Mathematics.

PROBLEM-SOLVING STRATEGIES

"Watch Me" – When a new lesson is introduced, this is the manner in which the author prefers to do it. The students sit in their desks, with their hands folded in their laps, simply watching what is being taught. seven examples are used to demonstrate the skill. In this manner, the students see seven examples of what the skill is, and how to apply the skill in order to solve the problems. They are not allowed to write anything down, while I am teaching the skill. If they ask a question, the response to them is to wait until the skill lesson is completed. After the time period, the students are then allowed to ask questions. They are constantly urged to denote the point where they no longer understood. For example, if Mary is asked, "what do you understand," an answer of "nothing" will not be allowed. By doing this, the students are enabled to think about where the understanding and the comprehension ended. It is believed that comprehension begins when the students are able to repeat what is said, as it is said. This activity may last 45 minutes. The author's normal class normally lasts between 75 and 90 minutes. The remainder of the period is spent allowing the students to answer questions, write down problems, and probing questions revealing those students who are still experiencing difficulties.

The second day of lesson introduction is primarily spent with students working problems on board on modeling the degree to which they understand. It is believed that areas of difficulty are visible at this time, thus relieving student stress. In order for students to solve problems on the board, they must feel that they will not be laughed at nor ridiculed for not understanding. The only way to display this is to have students go to the board. If a student laughs, the students are reprimanded by either removing him/her from the area and placing near the teacher, or the students' interruption is used as a discussion of what other ways are there for students to acknowledge their questions. There is also time for independent questioning, when students are working alone. It has been noticed that however, the more students that will orally discuss their concerns, the more student mastery occurs. Many students do need that independent help which could be addressed during the teacher planning/student tutorial time.

Another strategy used is student-to-student time. The students whom do understand assist those who do not. When pairing these students, take into account the willingness of the student to assist, and the willingness of the student to be helped. Monitor the pairs closely, and do not allow the assistance to continue more that 5-10 minutes. It is suggested that the student who has been helped be allowed to demonstrate his/her understanding. Many students will state that they cannot explain what they know. The preferred comment to them is if they cannot explain it, then it is a question of whether or not they understand it. The strategy previously mentioned normally takes place while the students are sitting at desks, writing on paper.

DIFFERENT WAYS OF CHECKING FOR UNDERSTANDING

Different ways of checking for understanding and completing student-to-student time are the use of long sheets of butcher paper or construction paper, and allowing the students to use markers or crayons instead of pencils. Students tend to enjoy the change of medium.

Different classes should be allowed to work together during student-to-student time. This skill fosters new relationships and a greater math understanding.

Checking for understanding is a daily activity. The more strategies that a student is given to solve a problem, the higher the level of understanding seems, and/or the quicker the understanding appears to arise.

Another problem-solving strategy that I use is called "problem dissection." The act of reading each sentence of a word problem that is at least three sentences or more. The author begins by reading the question of a word problem. It is usually the last sentence of a word problem. After the question is read, it is either highlighted or underlined. Then each sentence is read to determine whether the information in the sentence relates to the question, or contains information needed to solve the problem. The fewer sentences a student has to read, the more likely the problem will be solved accurately. When the student has finally solved a word problem, they write their answer as a sentence answering the question located at or near the end of the problem.

Test taking strategy – The students are urged to solve the computation or one-step problems first. This strategy may cause the student to feel empowered and armed with emotional energy needed to continue to solve the more difficult two and three step problems.

PERSONALIZED WORD PROBLEMS

Once it is believed or statistically shown that a majority of students have attained mastery of a particular skill, the students create or personalize a word problem. They have the option of changing an existing problem, or creating their one of their own. This is an excellent way of checking for mastery, and getting the students engaged and involved. A strategy for starting this is to write the first sentence on the board or ELMO. Then ask different students to add an additional sentence to the problem. Other students are constantly asked to decide and comment on the relation and reasonableness of each sentence to the problem. Begin with lesser difficulty and move on to problems with greater difficulty. Continue to encourage students, and set up parameters such as no curse words or drug paraphernalia terms used. The more advanced students may type their problems and email them if available.

CHARTS AND GRAPHS

LINE GRAPHS: CHARTING THE WEATHER

The teacher brings in poster size graph paper either 1-inch or 1 cm blocks. In the beginning of the instructional process, the teacher evenly places the days of the week and the numbers of degrees on the graph paper. The charting begins with the temperature at 6:00 am and ends with the end of the school day. Accu-Weather.com can be used to record the present temperature each hour. One student per class period is assigned to go to the site and record the temperature. Different colored stickers are placed near the board to represent each hour. Discussion should then ensue as to whether the temperature would rise or fall with variables, including season or month of the year, and climatic conditions. This is an excellent visual activity because the students can see the temperature changing over a period of time, and they are also using the internet as a research and teaching tool. Higher level thinking questions arise based on the temperatures for that day, such as what will be the temperature tomorrow? During the week of teaching of charts and graphs, these line graph posters should remain displayed in order for the students to be able to compare/contrast temperatures from one day to the next, or the entire weeks' temperature.

GRADE GRAPHING

A great way for students to keep up with their own grades is not only for them to record them, but to graph them. Under the premise that a line graph measures one thing changing over time, the students' grades are the items changing over time.

BAR GRAPHS

Human Bar Birthday Graph – The students line up in rows according to the month of their birthday beginning with January and ending with December. Depending on class size, the graph can be constructed as girls or boys only. Discussions should involve size of graph, comparing and contrasting the number of students in each month.

M&M® GRAPH

The students will graph the M&M candies in terms of color and number of each color on regular sheets of graph paper. This activity fosters an understanding of reasonableness. Teachers are to ask students questions such as how many candies should be in a fun-size bag of M&M's as opposed to a regular 10-ounce bag? Does the number of candies relate to whether the candies are the plain or peanut variety? If this activity is completed in groups, then the groups can compare the number of candies their individual group's bags of candies contain. Then when the groups relate their findings, the students should have a better understanding of the reasonable number of candies in a particular container of M&M candies.

CLASS TEST SCORES GRAPHS

Bar graphs are used to display individual class benchmark scores. Many students can see how their individual class scores relate to the other classes, and can serve as a gauge in attempting to raise the students' awareness, thus improving their benchmark scores.

Other such bar graphs that have been constructed include Boy/Girl graphs, pants/skirts, and uniform color top/bottom that have been done per class, and per team.

CIRCLE GRAPHS

Normally circle graphs have been completed based on percentages of items in a group. For example, finding the favorite color of students in a group. Either have them choose from a group of five items: for example, red, yellow, green, blue, or purple. Then tabulate the number of students that like a particular color, for example 4/20 of the students like red, then changing the fraction to a percent. Students must have an understanding of what a half, a third, and a fourth of a circle resembles. It is suggested that circles and percentages of circles are introduced and modeled frequently. In my experience, students seem to have an easier time reading and interpreting circle graphs rather than constructing them.

MEASUREMENT

In order to attain a greater sense of understanding from students in the component of measurement, they should use measurement tools. Rulers, paper clips, yard and meter sticks are basic tools for measuring length. Students tend to learn faster when using a ruler to measure the length and width of a book, the top of the desk, the chair, and the length from one classroom door to the next. Don't assume that the students know which end of the ruler to use. It is strongly suggested that use of measurement in the customary system is taught separately from measurement in the metric system. Once the skill is taught and modeled separately within the two systems, then review of both skills may ensue

Measurement of weight is also best when using tools that they can see and touch. Cups, pints, quarts, gallons, liters, and medicine droppers are just a few excellent tools to use when teaching these tools. Having students begin with the concrete and move to the abstract will foster a greater understanding of the skill.

When the teaching of measurement conversion begins, I have a particular system that has proven to be successful to my students, based on their benchmark scores. According to R. Charles et all (1999).

A problem such as:

    2 quarts = ____ pints

*Have them write the formula under the problem:

    2 quarts = ____ pints
    1 quart = 2 pints

*I Have them locate the 1 and cross it out. If the two numbers other than one are across from each other, as in the problem written above, the students will multiply.
In a problem such as:
   8 quarts = _____ gallons
*Have them write the formula under the problem:
   8 quarts = _____ gallons
   4 quarts = 1 gallon
*Have them locate the 1, and cross it out. If the two numbers are under each other, as they are in the problem written above, the students will divide.

The students should learn their measurement formulas for length, weight and time. The students will have access to a formula chart, yet they must understand which formulas to use with each problem. An abundance of modeling and guided practice is necessary for student success.

This skill must be taught and modeled over and over again, using whole group, peer grouping, and individualized instruction. Homework is necessary due to the fact that the students must practice on their own to determine where the difficulties lie. It is recommended that the chalkboard, and individual chalkboards be used.

FRACTIONS

REDUCING FRACTIONS

In order to successful in the reducing of fractions, it is believed that an introductory lesson of prime and composite numbers is necessary. Prime numbers are numbers that have two factors, one and itself. Composite numbers are numbers that have more than two factors. Begin with 1 and continue to 10. Tell the students not to ask any questions until they arrive at ten. This will give them an opportunity to listen and watch the procedure and allow for comprehension to take place. Volunteers are asked to tell whether the numbers from 11 to 20 are prime and composite based on definitions listed above. For numbers 30-50, mastery is constantly being determined. Then volunteers are asked to model procedures for determining whether the number is prime or composite. Students will be assigned to determine whether the numbers from 60 to 70 are prime or composite for homework. The next day's lesson begins with an oral review of definitions of prime and composite, accompanied by an abundance of modeling done by teacher and student. Working in groups, combined with individualized instruction will ensue for the next day. Once it is determined that a majority of the students are capable of determining prime and composite numbers, the lesson of reducing fractions begins.

The students are instructed to put the prime factors of the numerator and the denominator. For example:

$$\frac{2}{4} = \frac{2 \times 1}{2 \times 2} = \frac{1}{2}$$

The students will cross out the first 2 in the numerator and the first 2 located in the denominator. The students tend to understand this concept.

FRACTION MATCHES

Each student is given a card with a proper fraction on it. Using the skill of reducing fractions, students walk around the room looking for an equivalent fraction to theirs. Equivalent fractions are those in which the numerator and denominator can be either multiplied by the same number or divided by the same number.

ADDING AND SUBTRACTING LIKE FRACTIONS

An understanding of the definition of like fractions must begin this lesson. According to Charles et all (1999) like fractions are fractions that have the same denominator. For example:

$$\frac{1}{4} + \frac{2}{4} = \frac{3}{4} \qquad \frac{3}{4} - \frac{2}{4} = \frac{1}{4}$$

ADDING AND SUBTRACTING UNLIKE FRACTIONS

Unlike fractions are fractions that do not have the same denominator. For example:

$$\frac{1}{3} + \frac{1}{4}$$

The first step is finding the least common denominator of 3 and 4.

$$3 = 3, 6, 9, 12, 15, 18$$
$$4 = 4, 8, 12, 16, 20$$

The smallest or least common multiple (LCM) that 4 and 3 have in common is 12.

The second step is to multiply 1/3 by 4/4 as shown below:

$$\frac{1 \times 4}{} = 4 \qquad \frac{1 \times 3}{} = 3 \qquad 4 + 3 = 7$$

$3 \times 4 = 12$ $4 \times 3 = 12$ $12 + 12 = 12$
Using the like fractions from above:
$$\frac{4}{12} - \frac{3}{12} = \frac{1}{12}$$
(Charles et all, 1999)

CONCLUSION

Thus it is the responsibility of educators and policymakers in the educational arena to support and encourage research on innovative instruction to improve the quality of teaching and learning in K-12 schools so that all students will develop an appreciation and sincere interest in learning Math concepts and a joy for conquering the challenges provided by Mathematics. The afore-mentioned activities represent the numerous possibilities for improving the quality of delivering Mathematics lessons from an approach where children will enjoy the stimulation and sense of satisfaction when they receive high grades in Math as a result of their focus and concentration with the confidence that they will achieve and ultimately they do.

With the support of parents and community leaders, teachers will develop innovative Math learning activities that will help school-aged children K-12 grade to recognize how relevant mathematical concepts in their daily lives. All children need the opportunity to recognize their strengths as having a knowledge base to be academically successful in all subject areas, specifically Mathematics.

REFERENCES

AccuWeather.com. http://www.accuweather.com/.

Charles, R., Dossey, J., Leinwand, S., Seeley, C., Embse, C., (1999). Scott Foresman-Addison Wesley Middle School Math. Menlo Park, California.

Improving Student Achievement in Mathematics Part 1: Research Findings. http://www.apples4theteacher.com/resources/modules.php?op=modlad&name=News&file=article&sid=69&mode=thread&order=0&thold=0.

Iverson, D. (October, 2009). Using calculators in Elementary Math teaching. http://www.homeschool math.net/.../calculator-use-math-teaching.php.

Jeet, C. (October, 2008). Elementary Math Building Blocks for All Higher Mathematics. http://www. articlesbase.com/article-tags/elementary-math.

CHAPTER TWO
PREPARING STUDENTS FOR LEARNING: CONSIDERING THE EMOTIONS

MELBA VENISON, ROSE DUHON-SELLS, AND MATTIE SPEARS

Childhood and adolescent development is a difficult process under the best circumstances. The conflicts between the *id* (pleasure principle) and the *ego* (reality principle) present a constant struggle further complicated by the *superego* (moral and ethical restraints) as children mature. Today, children and adolescents attempting to engage in school in a meaningful way are not only confronted by this internal dilemma, but are also subjected to more life altering interruptions because of the enticing lures and effortless accessibility of the world. An additional aspect of this developmental stage is the *emotional roller coaster of* adolescence and the complex brain activity that is heightened during this period. Each decade seem to present a unique layer of difficulty for the child as the struggle to reach adulthood becomes more complex and unfortunately more tenuous.

Parents, teachers and administrators are usually the first to be affected by the development from the *terrible twos* to the *rebellious adolescent*. They are the usual targets for children and particularly adolescents as they attempt to establish themselves in their environment and in society. Normal development is characterized by rapid, orderly changes in multiple dimensions of functioning, including broad dimensions, such as cognitive, social, or emotional development and narrow dimensions such as expressions of anger or impulse control (Eyberg, Schuhmann, Rey, 1998).

The environmental system, according to A. J. Sameroff, includes basic life-sustaining habits such as eating, sleeping, and exercising and environmental hazards such as exposure to toxins, injury, disease, catastrophes and natural disasters. The exposure to toxins is of particular importance because this may have brain altering effects. The interpersonal environment of the child is influenced by the child to a similar extent as the child is influenced by them (Sameroff, 1995). The child makes mental and physical efforts to adjust to the demand of the environment and to structure it, and these efforts cause reorganization to occur with the child (Sameroff, 1975, 1995). Negative social activities render seemingly insurmountable obstacles for children as they struggled and continue to struggle with adjustment and adaptation for basic survival. It is imperative that the child makes mental and physical adjustments to the demands of the environment to structure it or be structured by it. If they fail, they are lost. Teachers and parents have the awesome responsibility to support, guide and develop an environment that is conducive for children affected negatively by the environments, as in the case of a broken home, loss of life or other such detrimental event.

This chapter focuses on strategies for teachers of children who exhibit an inability, for whatever reason, to *structure* the environment, and fall *victim* to its structure particularly when that structure does not represent a positive, nurturing, uplifting setting for normal growth and development. Unfortunately, this represents too many children sitting in classrooms oblivious to teachers who are baffled by the emotional and psychological roller coaster they see in the faces of their students. Many are not even aware of the image they project and have developed a hardened core that seems impenetrable. Victims of negative social and emotional events exhibit visages of confusion, pain, disbelief, and bewilderment. Many children have difficulty in focusing and are experiencing deprivation of basic needs as identified by Abraham Maslow. The first inclination may be to ignore them and work with those who are more responsive to instruction. There is an alternate possibility!

The information learned from the study of the brain offers a great deal of information relevant to the learning process. The brain is the only organ in the body that sculpts itself from its interactions with its environment. In a sense, our experience becomes biology. We used to think the brain you were born with was the brain you were stuck with, but we now know that learning experiences change and reorganize the brain's structure and physiology. The environmental factors mentioned earlier impact brain function considerably. For the mathematics teacher, any information regarding the student's background, family history, culture, perceptions, emotions, learning style and influences are of great significance when developing an effective academic program.

The following information has been written in the context of the teaching of literacy and mathematics. Consider the brain and what has been learned from the study of the brain. The brain is hard-wired to pay attention to and remember those experiences with an emotional component, whether it is the Challenger explosion or a particularly vivid simulation in which you took part in the eighth grade. However, emotional responses can have the opposite effect if situations contain elements that a person perceives to be threatening. In these situations, the amygdala starts a chain of physiological responses (commonly called the fight or flight response) to ready the body for action. Under these conditions, emotion is dominant over cognition and the rational/thinking part of the brain is less efficient (Woolfe, 2006).

The environment must be physically and psychologically safe for learning to occur. Brain research is relatively new and scientists agree that much is unknown about this complex organ, the core of bodily functions. There is, also, little consensus regarding the impact of the research findings on education; but researchers agree that the possibilities are tremendous. Eric Jensen (2000) asserts that there are important implications for learning, memory, and training. Research on the function of the brain, Jensen suggests, is being conducted by neuroscientists, researchers and educators in an effort to combine the findings of the brain/mind field with other fields to diversify and strengthen the applications. Neuroscience, though an important part of a larger puzzle, is not the only source of evidence. Neuroscience research combined with other fields like sociology, chemistry, anthropology, therapy and others, offer powerful applications for education. Neuroscience research is usually done at the molecular, genetic or cellular level where the concept of neurogenesis and the growth of stem cells are explored; while the applied cognitive sciences feature animal studies or clinical studies that show the real world behaviors which are more applicable in the education arena.

Important work in the area of neuropsychology involving brain research sheds light in better understanding of human brain functions. This work is so important that the United States' scientific community recognized the 1990s as the "decade of the brain" (Turgi, 1992).

Jean Piaget (1956) identified significant developmental milestones, which serve as a building block for this work. Scientists and education specialists continue to expand the understanding of the complex and integral processes of human brains, as we cognitively process life experiences.

Upon birth, each child's brain produces trillions more neurons and synapses than one needs to make connections in the brain, (Sousa, 1995) Researchers have identified a family of brain chemicals called neurotransmitters, which either excite or inhibit nerve cells referred to as neurons. Neurons have branches called dendrites to receive electrical impulses, which are transmitted through a long fiber termed an axon. The synapses between the dendrites join the process together and release neurotransmitters, which stimulate the neurons to collect and carry information for processing through a complex and systematic route. New experiences and information are filtered, then categorized contingent upon established brain structures as determined by prior knowledge and experience (Sousa, 2006).

Children are more adept at making new brain connections than are adults, and consequently they integrate new experiences at an incredibly fast rate (Newberger, 1997; Toepher, 1982). Therefore, rich learning environments yield more comprehensive brain pathways and proves to be advantageous for organizing and connecting meaning to learning experiences (Galloway et al, 1982). Systematic pathways facilitate the information processing and retention crucial to personal development. Opportunities of development learning most clearly occur when the brain demands certain types of input for stabilization of long-lasting structures in provision of organizational frameworks to retain future information. Commonly recognized milestones of motor development, emotional control, and vocabulary development are key indicators in sequentially formative development progression. A normally functioning brain has the learning readiness to receive the information necessary for each skill acquisition.

A study by Paul Thompson et al (2000) reported that optimization of brain connections occurs early in childhood as well as just before puberty. Although the brain of a six-year-old child has grown to 95 percent of the adult brain size, size is not as important a factor as putting to use the brain cells that are produced. Because a second wave of brain cell production takes place just before puberty, it is important that young people are encouraged to take advantage of optimizing their brain activity by becoming involved in reading, physical activity, and musical skill development during elementary school age. Musicians and athletes are often most successful when they begin their training at a very young age; avid young readers continue to read throughout adulthood. These are lasting activities that "wire" the brain for use later in life. The brain connections that are used will survive and those that are not used will not survive. It is also important to realize that drugs and alcohol have devastating effects of the brain connections in the pre-teen years when the brain is still in developmental stages.

Though the concept of brain-based learning excites many in the field, the concept is not without critics. Critics worry that so much is unknown about the brain and its function; it is ludicrous to make assumptions based on presumptions rather than scientific evidence. Unfortunately, many of the myths about the brain are professed by consultants and classroom teachers who misinterpret or misunderstand the findings of the studies. One of the most vocal critics is Dr. John T. Bruer, *Kappan* (Phi Delta Kappa), who writes,

> These issues ["The American School Board Journal" (February 1997), "Educational Leadership" (March 1997), "The School Administrator" (January 1998), and now "The NASSP Bulletin" (May 1998) address the implications of the new brain research for educators.] contain a variety of articles – articles by advocates of brain-based curricula, articles by educational futurists, articles by cognitive (not brain) scientists. In fact, *it is rare to find an article written by a neuroscientist in the educational literature*. Of these articles, those citing cognitive research – provide the most useful advice to educators. Educators should be aware that cognitive science – the behavioral science of the mind – is not the same as neuroscience – the biological science of the brain.... *Most other claims found in the emerging brain and education literatures are vague, outdate, metaphorical, or based on misconceptions.*

Nevertheless, there is sufficient evidence to support the benefits of the research and the scholarly discussions about the brain and learning.

Advocates are undaunted by the criticism. Many suggest that although much has yet to be learned about the brain, what *is* known is helpful as the research continues to unfold new knowledge about the brain, its development and its function. David Sousa and Eric Jensen agree and point to recent research which has provided valuable information for educators who develop curriculum for early learners.

According to Jensen "normal childhood experiences usually produce normal kids, but and there are *'windows'* of opportunity for development." Some must be adhered to and others can be delayed. The most critical "windows" are those for our senses, the *parent-infant emotional attunement, language learning* and *non-distressed sense of safety*. These "windows" are time-sensitive and cannot be recaptured. Jensen indicates that learning is strengthened in the brain through repetition, but boredom damages the process. Additionally, research indicates that many factors influence learner success including parents, peers, genes, trauma, nutrition and environment, but there is no way to quantify them. The human brain can and does grow new cells and these cells do interact with existing cells. Current research is exploring the interaction of the new and old cells and how that information can add to the body of knowledge about how learning occurs.

Jensen summarizes that brain-based learning is not a panacea or magic bullet to solve all of education's problems, but there are numerous examples of practices that have changed or have been impacted by what has been learned. He cites the work of Dr. Michael Merzenich (Stanford University) and Dr. Paula Tallal (Rutgers), two neuroscientists who developed the phonological processing improvement product, FastForword that uses discoveries in neural plasticity to change the brain's ability to auditorally process the printed word.

Another area explored by neuroscience is the growing knowledge about the development of emotional and social competence. Increased interest in emotional and social competence was motivated by Daniel Goldman (1995) who defines Emotional Intelligence as a form of intelligence relating to the emotional side of life, such as the ability to recognize and manage one's own and others' emotions, to motivate oneself and restrain impulses, and to handle interpersonal relationships effectively (Goldman, 1999).

The early work of developmental psychologists, like Piaget, who focused on cognitive development, set the stage for the current research agenda on social and emotional development in children. But in recent years child development research has expanded its scope to study the unfolding of children's social and emotional lives. The result has been a new understanding of what, for example, makes a child socially adept or better able to regulate emotional distress. This new strata of scientific understanding can be of immense help in informing the practical efforts of the emotional literacy movement (Salovey and Sluyter, 1997). Research in this area has resulted in a clearer understanding of a child's social adaptations, ability to monitor emotions and cope with emotional distress. This new information serves as a basis for teaching reading to students who find difficulty in interacting with the literature on any level due to prior emotional experiences that may have a negative impact on normal development.

The brain, which controls the emotions and the impetus for social interaction is the last organ of the body to mature anatomically, and continues to grow and shape itself throughout childhood into adolescence. Similarly, as the development progresses, circuits that control emotional competence seem to be the last part of the brain to mature. Because of the plasticity of the brain it can reshape itself as it responds to repeated experiences that strengthen the circuits. Children are more adept at making new brain connections than are adults, and consequently, they integrate new experiences at an incredibly fast rate (Newberger, 1997; Toepher, 1982). Therefore, a rich learning environment yields more comprehensive brain pathways proven to be advantageous for organizing and connecting meaning to learning experiences (Galloway, 1982; Newberger, 1997; Sousa, 1995). What this means is that childhood is the most optimum time to provide children with repeated experiences that help them to develop healthy emotional habits – for self-awareness and self-regulation, for empathy and social skill.

Children's emotions must be recognized and their importance for learning accepted. Studies by R. Sylwester (1994) reveal that emotions drive attention and, in turn, drive learning and memory. Since more neural fibers project from the brain's emotional center into the logical/rational center than the reverse, emotion tends to determine behavior more powerfully than rational processes. Thinking is an integrated process of the body/brain system. According to Sylwester the emotional system is located in the brain, endocrine, and immune systems and affects all other organs of the body. Therefore, chronic emotional stress has adverse affects on the entire body. Stressful school environments inhibit learning while positive classroom atmospheres encourage chemical responses in students that help them learn. Children naturally seek out and thrive in places where caring is present. Integrating emotional expression in a caring classroom atmosphere improves memory and stimulates the brain to learn.

These findings have implications for children who are victims of a culture and an environment often violent, hostile, and threatening. They present unique challenges for teachers who leave Teacher Education Programs armed with theories and learning models that often do not take into consideration the impact of negative experiences on the brain and its development. Previously taught theories have proved to be effective with students who have normal developmental experiences, but for those who are faced with violence, negligible self-worth, and neglect from birth are less responsive and often stagnate in their developmental processes.

Sylwester believes curricula must include many sensory, cultural, and problem layers that stimulate the brain's neural networks. He calls for classrooms that are closely related to real-world environments. Current studies support the critical role adults play in facilitating an early stimulating environment for children, but if caregivers are absent or preoccupied with the necessities of life such as food and shelter, and are unable to provide early stimulation, students stagnate and brain activity is limited at a time when children are most receptive to stimulation.

In response to the growing number of students who are demonstrating an inability to control emotions, school-based programs have surfaced across the country in an effort to address the competencies in emotional and social development. These programs have a variety of names from "social development," "life skills," "self-science," and "social competency", but basically, they all address the same issues in providing students with coping mechanism which are socially acceptable and self-fulfilling (Salovey, 1997). The Collaborative for Social and Emotional Learning, begun at the Yale Child Studies Center, in New Haven, Connecticut, has provided school districts with quality programs that offer practical assistance in this area. The Nueva School in Hillsborough, California, was the first to start such a program, and New Haven was the first city to implement such a program in public schools district wide.

The function of the brain is directly impacted by emotions. When teachers understand child and adolescent development and focus on strategies to validate children in a way that will invite them to be a part of the greater body, the brain becomes receptive to new information and new learning.

Without doubt, the task confronting reading teachers is daunting. Fortunately, teaching children to develop a comfort level with mathematics is a mandate we cannot afford to ignore. With the proliferation of data about how to teach children and increase proficiency in mathematics, there still exist gaps between children with rich mathematical exposure and those with little or no early exposure. It is encumbered upon the parents and teachers to forge an alliance that will support, nurture and validate children so they can freely exposed their minds to the wonders that await them. Together we must remove the barriers that impede them from experiencing a happy, content and productive life. Learning to integrate mathematics skills into daily life is the key to that life!

REFERENCES

Galloway, L. (1982, Spring). "Bilingualism: neuropsychological considerations." *Journal of Research and Development in Education*, 15, 12-28.

Goldman, D. (1995) *Emotional intelligence.* New York: Bantam Books.

Jensen, E. (2000) *Brain-based learning.* San Diego: Brain Store Incorporated.

Newberger, J. (1997). "New brain development research: A wonderful window of opportunity to build public support for early childhood education!" *Young Children*, 52, 4-9.

Piaget, J. (1954). *The construction of reality in the child.* New York: Basic Books.

Salovey, P. & Sluyter, D. (Eds.). (1997). *Emotional development and emotional intelligence: Implications for educators.* New York: Basic Books.

Sameroff, A. J. (1995). "General Systems theories and developmental psychopathology." *Development Psychopathology,* 1, 659-695.

Sameroff, A. J. (1975). "Early influences on development: Fact or fancy?" *Merrill-Palmer Quarterly*, 21, 267-294.

Sousa, D. (2006). *How the brain learn mathematics.* Reston, VA: The National Association of Secondary School Principals.

Sylwester, R. (2003) *A biological brain in a cultural classroom: enhancing cognitive and social development through collaborative classroom management.* Thousand Oaks, CA: Corwin Press Incorporated.

Thompson, P. M., Giedd, J. N., Woods, R., MacDonald, D., Evans, A., and Toga, A. (2000). "Growth patterns in the developing brain detected by using continuum mechanical tensor maps." *Nature*, 404, 190-193.

Toepher, Jr., C. (1982). "Curriculum design and neuro-psychological development." *Journal of Research and Development in Education*, 15, 1-10.

Turgi, P. (1992). "Children's rights in America: The needs and the actions." *Journal of Humanistic Education and Development*, 31, 52-63.

CHAPTER THREE
FACTORS INFLUENCING STUDENT ATTITUDES, BEHAVIORS, AND LEARNING OUTCOMES IN THE STATISTICS COURSE

LISA EARGLE, ASHRAF ESMAIL, AND JAS SULLIVAN

INTRODUCTION

Many studies have addressed the problems in teaching math to students who struggle in learning math in PK-12. However, a small number of studies have looked at improving mathematics ability for university level students. One area of mathematics that students have trouble with is statistics. The objective of this study is to analyze findings from research conducted at the university level to improve achievement in learning statistics. This paper addresses the question of what factors influence student attitudes, behaviors, and learning outcomes in the statistics course.

LITERATURE REVIEW

According to Gregory Byrne (1989),

> most Americans students leave school with such a poor understanding of mathematics that they cannot adequately perform the vast majority of jobs, much less consider specialized careers in mathematics or science (p. 597).

This could be caused by a number of reasons. William Nibbelink (1990) argues that we need to introduce introductory work with equations at the elementary level which will allow students to perform better at the high school level. Other research has indicated that anxiety could be at play here. According to Mark Ashcraft and Jeremy Krause (2007)

> standardized math achievement test varies as a function of math anxiety, and that math anxiety compromises the functioning of the working memory. High math anxiety works much like a dual task setting: Preoccupations with one's math fears and anxieties functions like a resource-demanding secondary task (p. 243).

Research has looked at number of strategies of improving mathematical achievement of students experiencing mathematics difficulty. Scott Baker, Russell Gersten, and Dae-Sik Lee (2002) indentified a number of interventions including: a) providing teachers and students with data on student performance; b) using peers as tutors or instructional guides; c) providing clear, specific feedback to parents on their children's mathematics success; and 4) using principles of explicit instruction in teaching math concepts and procedures (p. 51). Linda Proudfit (1992) suggests that using effective questioning techniques helps students ask themselves questions to assist them through the problem-solving process. According to Proudfit (1992),

> students who acquire the habit of asking themselves questions may not need to ask many questions of others. This should help students develop confidence in determining the correctness of their solutions and in becoming independent thinkers (p. 135).

James Lalley and Robert Miller (2006) conducted a study comparing and examining the impact of pre-teaching and re-teaching on academic self-concept and math achievement of third grade students identified as low achievers. Results showed that pre-teaching and re-teaching drew significant increases in math problems, math computation, and math concepts. Thomas Slater, Hazel Coltharp, and Steven Scott (1998) found that lead teachers who developed hands-on-manipulative workshops through extensive use of video-teleconferencing technology made a positive impact on the quantity and quality of integrated math instruction. Finally, David Green (2002) found that offering lesson plans using math games that utilize cooperative and creative problem solving can be useful for children struggling in math. The game employed many of the NCTM Content standards as well as a range of problem-solving methods. It allowed students to make discoveries as a class, in small groups, and independently.

Other research has looked at techniques for teaching math with students with learning disabilities. Cynthia Wilson and Paul Sindelar (1991) conducted a study looking at the impact of strategy sequencing and teaching practice problems in teaching students with learning disabilities to indentify the correct algorithm for solving addition and subtraction word problems. Results found that students in the strategy-plus-sequence group, as well as those in the strategy-only group, scored significantly higher than did students in the sequence-only group. Findings showed that strategy teaching was the most effective of the two instructional components. Bottge et al. (2007) found that students with learning disabilities benefited from multimedia and hands-on formats in improving math achievement. Brian Bottge and Ted Hasselbring (1993) studied two groups of adolescents with learning difficulties with mathematics and compared them on their ability to generate solutions to a contextualized problem after being taught problem-solving skills under two conditions, one involving a contextualized problem on videodisc, the other one utilized standard word problems. The problems focused on adding and subtracting fractions in relation to linear measurement and money. Results showed "that students with a history of learning difficulties with a history of learning difficulties can learn how to solve complex, meaningful mathematics and, further, can transfer these skills to other real-world tasks" (p. 564).

> Both groups of students improved their performance on solving word problems, but students in the contextualized problem group did significantly better on the contextualized problem posttest and were

able to use their skills in two transfer tasks that followed instruction (p. 556).

At the university level, one study finds that women perform much better on math problems when testing in a setting without men (Anonymous, 2000). The study found that "simply being in a classroom with men did not affect women's overall intellectual performance... The performance differences were limited to situations in which women were tested on math, a subject that is traditionally stereotyped" (p. 154).

Much of the literature has focused on how mathematical skills can be improved at the PK-12 level. Previous research has not looked at the effects of instruction on achievement of students in a statistics who are at risk of failure. The focus of this paper is to identify techniques that can assist students at the university level improve mathematical and more specifically statistical skills.

DATA AND METHODS

The data for this paper are based upon the authors' observations made in the classroom, from teaching data analysis, statistics, and other quantitative methods-based courses, for over a decade at different universities in the southern and mid-western United States. The courses in which the authors' taught were all upper-division (junior and senior-level) courses that mostly social science and humanities majors were enrolled. Most of the students enrolled in these courses, as well as the universities, were first-generation racial and ethnic minority students from poverty or working class backgrounds. The authors of this paper (who were the instructors in these courses) are racial and ethnic minorities as well. One of the authors is a first generation college student from a working class background.

FINDINGS

As shown in Figure 3.1, there are four categories of factors, within the course environment, that influence student behaviors and activities. There are (1) procedures used by the instructor to teach the course material; (2) the instructor's disposition, attitudes, and behavior; (3) the classroom's physical environment; and (4) other student, instructor, and course features. In the sections below, we address each of these factors. These factors, in turn, affect student performance.

FACTORS INFLUENCING STUDENT LEARNING OUTCOMES

Instructional Procedures------------------

Instructor Disposition-------------------- Student   Learning
                                         ------Behavior &-----→Outcomes
                                              & Attitudes

Classroom Environment------------------

Other Features----------------------------

INSTRUCTIONAL PROCEDURES

SETTING THE STAGE FOR THE COURSE

On the first day of class in the course, it is very important to provide students with a detailed syllabus. The syllabus sets the tone for the rest of the course and it also serves as an important reference for students. A syllabus should contain several items: the name of the course; the course location and meeting times; instructor's name and contact information; required textbooks; an overview of the course, course goals and objectives; description of course requirements/assignments, their due dates, and their contribution to the final course grade; and a schedule of course topics, associated readings, and their class coverage dates. The instructor should not only distribute paper copies to the students in class, in person, but also have an electronic copy available for students to access. Furthermore, it is useful to lecture on the syllabus, to be sure to address any questions that the students may have about the course and to be able to elaborate further on specific items.

Another important procedure is taking attendance daily in class, including all exam days. This indicates to students how imperative it is for them to attend class every day. In courses such as statistics or math, missing one class day can mean the difference between passing and failing an exam. For students to adequately comprehend the material that is addressed in a statistic's course, most students need to observe, first-hand, the course lectures. Most students cannot discern, simply from reading assigned material in a textbook and/or acquiring a copy of the class notes from another student, the process involved in completing a statistical or math problems and the reasoning behind the interpretation of the problems' results.

Likewise, it is important for the instructor to arrive at class on time and prepared for the class. This emphasizes to students the seriousness of the course. Moreover, the instructor serves as a role model for students, demonstrating to them that they should arrive to class on time and prepared to listen and learn. Students closely observe their instructors' actions and adjust their behaviors accordingly. They also tell other students about their instructors, so new students will enter an instructor's class with a set of expectations about the instructor and course before the new semester/course. If these expectations are negative, the instructor will have a more difficult or challenging time teaching the course because old expectations will have to be erased and replaced with new expectations. Moreover, if possible, students will avoid instructors with the reputation of arriving in class unprepared and the instructor's enrollments will suffer.

A fifth important procedure to initiate in the first class meeting is the regular asking of students if they have any questions. Students may be very reluctant, without prompting by the instructor, to ask questions for the fear of appearing stupid. By asking the students if they have any questions, the instructor is indicating to them that having questions is expected and that the instructor is receptive to answering questions.

Students also need to be reminded, occasionally, that the instructor is available to answer questions outside of class during regular scheduled office hours or otherwise. The authors of this chapter have both regularly scheduled office hours and an open door policy. If their office doors are open, students are welcome to ask questions even if the time is outside of their regular office hours. We also begin each meeting of our classes by addressing any questions students may have about the homework, projects, or other assignments that they are completing for the course.

Moreover, if necessary, the instructor should offer extra review or other organized study sessions outside of class. Having this extra, not required of the instructor by the school administration, support system in place makes many students feel more at ease with the course. Also, if the instructor provides opportunities for students to work together in small groups, whether in extra study sessions or homework assignments, students do not feel as isolated and alone in facing a sometimes difficult course.

Ninth, instructors need to submit textbook orders to campus bookstores by the deadlines, to aid the bookstore in obtaining textbooks before the new semester begins. Instructors should also check with the bookstores after submitting the purchasing orders to make sure that the textbooks requested are available and that enough copies have been obtained for students to purchase. It is also helpful if the instructor can inform students, on the first day of class, about the cost of the course textbook(s).

Tenth, numerous homework, examinations, and other assignments should be given throughout the semester, so that no one grade will determine a student's performance in the course. Often students will be forced to manage multiple deadlines and tasks in the semester and may not consistently put forth their best efforts in the course. On the other hand, these numerous assignments will keep the statistics course foremost in their thoughts and actions.

COURSE LECTURE PROCEDURES

Students differ in terms of their learning styles, with some quickly absorbing information from seeing it, others from hearing it, and still others need to physically manipulate objects. Multiple approaches to addressing the material can easily be incorporated into a statistics or mathematics course. One approach that instructors can use is to write lecture material on the board and require students write the material down as notes. This will aid student memorization of the material. When working through a problem, the instructor should show every step in the problem and label all parts of the problem. Students should be required to do the same in completing their homework, exams, and other assignments.

As the instructor works through the problem, he/she can verbally tell the students what he/she is doing and provide explanations for it. The instructor should be sure to speak loudly enough for students to hear him/her and to enunciate words clearly enough for students to understand what he/she is saying. While verbally addressing the problem, analogies can be incorporated to further clarify statistical concepts. For example, an analogy of slicing a pie into equal parts or a river having two banks can be used to convey the concept of class boundaries in statistics class. To make the class more interesting and to promote student involvement, the instructor can ask students to provide other examples/analogies to illustrate a particular concept. Analogies not only make the material interesting, but aids students in relating statistical concepts to everyday life. The instructor can advance this approach further by using data from everyday life (such as crime rates of cities, size of students' families, or types of jobs) to calculate statistics such as means, standard deviations, or correlations.

Moreover, to promote active student involvement in the class, once the instructor has worked through at least one example of a problem, he/she can ask students to "help" him/her complete another example problem. Repetition of material is key in a course such as statistics, where students are already intimidated by numbers. The more examples students are provided, especially ones in which they were active participants, encourages students to take "ownership" of the course and knowledge that is produced within it. This empowers students, providing them with the self confidence needed to master difficult material.

Providing multiple examples of a concept or problem in class additionally limits the volume of material or concepts that can be covered within one class meeting. Covering many ideas within one class meeting, while physically feasible to do, can easily overwhelm students.

Another visual approach to teaching statistics involved the use of graphs and other visual display techniques to convey concepts. For example, producing a scatter plot and showing co-variation as part of a lecture on correlation or regression analysis demonstrates to students the basis for these types of statistical techniques. Having students work with physical objects (such as counting the number of chairs in classrooms and calculating the mean or counting the number of cars by color category and finding the relative percents) can aid students in understanding course material.

It is useful for students to write, in words, the meaning of symbols used for statistical concepts and calculations. For example, when showing students for the first time the formula for calculating a sample mean, it helps students understand the statistical procedure if the instructor writes out in words, "the mean is equal to the sum of each value, times its frequency, then divided by the sample's size." Along with writing out in words the meaning of symbols and formulas, instructors should limit the use of jargon to the minimum required to adequately teach the course. Excessive jargon or use of "big words" confuses students more than it does to increase their vocabularies.

Another strategy that instructors can use involves anticipating student questions and points of confusion that may arise during the lecture and deciding ahead of time, how he/she will address those issues. For example, probability is one topic within the statistics course that many students find difficult. However, if the instructor can relate that concept (probability distribution) to another concept (relative frequencies) covered earlier in the course, students are more likely to understand the concept.

INSTRUCTOR'S DISPOSITION, ATTITUDE, AND BEHAVIOR

Students may initially show little enthusiasm for a course involving numbers. Hence, it is important for the instructor, from the first day of the class until the last day of class, to show interest and enthusiasm for the material he/she is teaching. One way to do this is to incorporate into the course lectures and assignments topics in which the instructor conducting research or personally interested.

Also, it has been the experience of the authors that student performance in the statistics course can be inconsistent. This often frustrates the students themselves. To encourage students to continue working with the material and staying engaged in the course, the instructor needs to maintain a positive attitude about the students' capabilities. Offering occasional words of encourage to students, while infrequently criticizing their performance, is useful for stimulating student engagement in the course. Remaining calm when students are upset and frustrated reassures them that they are going to be okay.

Third, the instructor should not assume that something is so easy or is common knowledge that everyone should know it. If a student asks a question about something, regardless of how mundane it is, the instructor should answer it and discourage any would-be mocking classmates from making rude comments. What the instructor wants to foster in the course is an attitude of respect and appreciativeness, for both the instructor and student enrolled in the course.

CLASSROOM PHYSICAL ENVIRONMENT

There are many important features of a classroom's physical environment that can impact student learning. These features include: (1) lighting; (2) temperature; (3) seating/desks; (4) view of the blackboard; (5) blackboard space; (6) outside noises; (7) acoustics; (8) spacing between desks; and (9) availability of audio/visual aid equipment.

If the lighting is too bright or dim in a classroom, students may have difficulty focusing on the material because they cannot see the board or develop headaches. If the instructor has any power to decide which classrooms he/she can use, he/she should find one with appropriate lighting. Likewise, a classroom that is too hot or too cold acts as a distraction for both the students and the instructor. It is difficult to spend the necessary amount of time covering material in class if one is freezing or sweating.

Third, if possible, an instructor should make sure that there is adequate seating for the students enrolled in the course and that the chairs/desks are in good condition. No one wants to sit in a chair and be concerned as to whether it will hold him/her for the duration of the class or drop him/her on the floor. If the instructor can choose a classroom with wide, comfortable chairs, he/she should use it. Not all students are small in size or enjoy sitting on a metal seat for an hour. Also, it is preferable to use a classroom where there is adequate spacing between desks, so that the instructor can easily move around the classroom to answer students' questions, to proctor examinations, and students have some personal space.

Fifth, there should be adequate board space to use for the course, so that the instructor can spend his/her time teaching, not erasing material. He/she should also seek classrooms where the students' view of the board is not obstructed by podiums or equipment. Sixth, if possible, classrooms where is the acoustics within the classroom are adequate and outside noises do not penetrate the room should be selected for the course meetings.

ADDITIONAL FACTORS

Additional factors that influence student learning are characteristics of students enrolled in the course, instructor characteristics, and other course features.

STUDENT CHARACTERISTICS

Some student characteristics that research has found to impact performance are (1) the number of students enrolled in the course; (2) the racial/ethnic composition; (3) gender ratio; (4) socioeconomic class; (5) class rank/age of students enrolled in the course; and (6) percentage of students repeating the course. It is helpful for the instructor to realize that students enrolled in larger courses and that are racial and ethnic minorities, female, economically disadvantaged, in their first year of high school or college, and/or repeating the course are less likely to perform well in the course. If possible, the instructor should provide extra help to these students as warranted.

INSTRUCTOR CHARACTERISTICS

Some instructor characteristics that can influence student learning in the course are: (1) experience teaching the course; (2) race/ethnicity; (3) gender; (4) age; (5) was first generation college student or from disadvantaged background; (6) instructor characteristics match-up or differ from student characteristics. Instructors with years of experience are familiar with the issues involved in teaching and are more equipped to address those issues adequately. Instructors that are racial and ethnic minorities, female, and/or first generation college graduates – especially if they are teaching students with similar characteristics – can promote student learning by inspiring the students as a role model.

OTHER COURSE FEATURES

Other course features that may impact student learning are: (1) course meeting length; (2) how often course meets; (3) time of day for course meetings; and (4) how integrated course is into the rest of the curriculum. Long class periods (exceeding an hour) require that the instructor give a short break for trips to the restroom or vending machines or to just move around. The more frequently a course meets, the easier it is for students to retain the material covered in a previous class meeting. Classes that meet very early or late may have students that are tired and having difficulty focusing on the course material. Finally, if the course is well integrated into the rest of the curriculum, many of the concepts cover in the class will be addressed in another course. This will help students retain the information that they learned in the statistics course. It also makes the students see the course as being relevant, because they will have to use this knowledge from the course repeatedly. Hence, they might as well learn as much as they can in the course or suffer future difficulties in other courses.

CONCLUSION

There are many factors which can impact student behaviors and attitudes in the statistics or math-based course and ultimately, affect student learning outcomes. Some of these factors are under the control of the instructor, such as teaching strategies and instructor attitudes. Other factors, such as the classroom's physical environment and the characteristics of students enrolled in the course, the instructor has limited or no ability to change. Attention to these factors and their implications for teaching the course is especially important in a number-based course, where typical students are ill prepared to perform. These factors are even more salient in a course where students are arriving from disadvantaged backgrounds, because of the tendency for those students to receive preparation from disadvantaged schools, where teachers have few resources and less preparation to instruct their students. Coming from a disadvantaged school can have accumulating impacts upon students' future performance in courses and advancement in their educations (Schaefer, 2006).

REFERENCES

Anonymous. 2000. "Women's Math Performance Improves When Tested Without Men." *Journal of College Science Teaching*, 30, 3, 154.

Ashcraft, Mark H. and Jeremy A. Krause. 2007. "Working Memory, Math Performance, and Math Anxiety." *Psychonomic Bulletin & Review*. 14, 2, 243-248.

Baker, Scott, Russell Gersten, and Dae-Sik Lee. 2002. "A Synthesis of Empirical Research on Teaching Mathematics to Low-Achieving Students." *The Elementary School Journal*, 103, 1, 51-73.

Bottge, Brian A., Enrique Rueda, Ronald C. Serlin, Ya-Hui Hung, and Jung Min Kwon. 2007. "Shrinking Achievement Differences With Anchored Math Problems," *The Journal of Special Education*, Vol. 41, No. 1, 31-49.

Bottge, Brian A., and Ted Hasselbring. 1993. "A Comparison of Two Approaches for Teaching Complex, Authentic Mathematics Problems to Adolescents in Remedial Math Classes." *Exceptional Children*, 59, 6, 556-566.

Byrne, Gregory. 1989. "Overhaul Urged for Math Teaching." *Science*, 243, 597.

Green, David A. 2002. "Last One Standing: Creative, Cooperative Problem Solving." *Teaching Children Mathematics*, 9, 3, 134.

Lalley, James P. and Robert H. Miller. 2006. Effects of Pre-Teaching on Math Achievement and Academic Self-Concept of Students with Low Achievement in Math." *Education*, 126, 4, 747-755.

Nibbelink, William H. 1990. "Teaching Equations." *The Arithmetic Teacher*, 38, 3, 48-51.

Proudfit, Linda. 1992. "Questioning in the Elementary Mathematics Classroom." *School Science and Mathematics*. 92, 3, 133-135.

Schaefer, Richard T. 2006. *Racial and Ethnic Groups*. Tenth Edition. Upper Saddle River: Prentice Hall.

Slater, Thomas F., Hazel Coltharp, and Steven A. Scott. 1998. "A Telecommunications Project to Empower Kansas Elementary/Middle Level Teachers as Change Agents For Integrated Science and Mathematics Education." *School Science and Mathematics*, 98, 2, 61.

Wilson, Cynthia and Paul T. Sindelar. 1991. "Direct Instruction in Math Word Problems: Students With Learning Disabilities." *Teaching Exceptional Children*, 57, 6, 512-519.

CHAPTER FOUR
MULTICULTURAL EDUCATION MATH STRATEGIES FOR TODAY'S DIVERSE CLASSROOMS

ALICE DUHON-McCALLUM, DORIS L. TERELL, AND GWENDOLYN DUHON

The purpose of this chapter is to provide multicultural teaching strategies for those who teach math in today's increasingly diverse classrooms. According to Duhon-Sells (1992), the multicultural thrust should be implemented so that teachers can learn to utilize the unique cultural assets of their students in creating a learning environment that will be beneficial to all students, regardless of their background.

The multicultural education movement is the result of national awareness of the academic needs of students in this culturally diverse society. During the past five decades there has been much discussion in reference to an effective teacher education curriculum that prepares teaches to do more than tolerate the cultural differences among their students. In response to the ongoing revitalization movements, educators, social activists, civil righters, and human dignity supporters tapped into the moral power prompted by a 300 year history of victimization and oppression and began to become inspired thus starting an implementing force toward implementing programs, and teaching practices. Theoretical and academic interest groups began to respond more adequately to the learning and development of ethnic, cultural and racial groups, both urban and rural, in an attempt to help them become more structurally integrated into our societal fabric.

When teachers develop a positive attitude about multiculturalism and incorporate it in their curriculum, students of all cultural backgrounds will be successful. Multicultural promotes a positive change for persons of all cultures. It not only involves teaching majority groups about minorities, but it also, teaches minority groups about the majority groups. If properly incorporated into the curricula, multiculturalism will unite our divided nation into one unit that has no mainstream cultural but many diverse subcultures that will work together for the good of all cultures and not just the culture of the majority or minority.

Many high school students were never taught from a perspective that included views of different cultures. Some students acknowledged that what was taught in their high schools was taught from a middle class, white perspective. If students of color are not taught about themselves, their collective cultures, and their surroundings, they cannot be expected to survive in their society. To survive, students must be taught using multicultural concepts to help them to grow and become productive citizens within American society. This survival technique begins with education. Education begins at home and educators continue the process in the classroom.

Multicultural mathematics is the application of mathematical ideas to problems that confronted people in the past and that are encountered in present contemporary culture. In attempting to create and integrate, multicultural mathematics materials related to different cultures and draw on students own experiences into the regular instructional mathematics curriculum, WG-10 on Multicultural/Multilingual classrooms at ICME-7, in 1992 pointed out aspects of multicultural mathematics:

An aspect of multicultural mathematics is the historical development of mathematics in different cultures (e.g. the Mayan numeration system). Another aspect could be prominent people in different cultures that use mathematics (e.g. an African-American biologist, an Asian-American athlete). Mathematical applications can be made in cultural contexts (e.g. using fractions in food recipes from different cultures). Social issues can be addressed via mathematics applications (e.g. use statistics to analyze demographic data (p.3-4).

The growing body of literature on multicultural education is stimulated by concerns for equity, equality, and excellence as part of a context of diversity. Teachers realize that students become motivated when they are involved in their own learning. This is especially true when dealing directly with issues of greatest concern to themselves (Freire, 1970).

Educators of mathematics must stand together to dismantle many of our student's archaic mathematical phobias and reverse these phobias into visions of academic success in math. Today, the number of students taking mathematics is increasing. Teaching strategies to help minimize some of the common mathematics phobias and difficulties students have can positively impact students. Mathematics teachers can play a greater role in multicultural education by developing and implementing strategies to assist student in the understanding of mathematics in a culturally diverse learning environment. C. E. Sleeter and C. A. Grant (1987) stated that multicultural education approaches should generally on concentrate on developing curriculum and instruction practices. Until multicultural approaches become a common place in our post-secondary institutions, individual educators and their classrooms must continue to practice a multicultural approach to their teaching whenever applicable.

A continuous and concerted effort to use various motivational techniques needs to consider in the instruction of mathematics, especially for students in special education and remedial mathematics classes. The teaching of mathematical principles by a appealing to the sensibilities of the students can be more effective and productive (NCTM, 1991).

Like students, teachers have been blamed for their failure to meet the challenge of diversity in the classroom. While teachers' attitudes, idiosyncrasies, biases, prejudices, and perceptions do influence the success of their student (Ladson-Billigns, 1994. Bennett, 1995; Niet, 1996), it is unwise to use them as scapegoats in an attempt to explain school failure. Generally, teachers tend to be the product of the teacher-preparations programs they were in. If these programs lack adequate ingredients for preparing prospective teachers, then these teachers will reflect to inadequacy of the preparations process. Multicultural education is also a source empowerment for teachers. To do so, multicultural programs have a set of goals for teachers that are equally important. Accordingly, math teachers must:

1. Know, understand and appreciate difference math experiences and contributions of minorities and other ethnic groups in American society.
2. Show a thorough understanding of the nature of the pluralistic society and the educational implications of diversity in math education.

3. Develop a sound rationale of multicultural math education through a philosophical base that incorporates pedagogical principles that can be transferred to curricular areas;
4. Enhance the optimality of academic and social development o their students through the knowledge and process having sociocultural factors that influence the learning process of math and science;
5. Understand students' attitudes, values, and other motivational forces that affect the performance and achievement of students in math lessons;
6. Acquire knowledge in math pedagogical techniques that are pluralistic and multicultural.
7. Utilize multicultural math materials that are sensitive and relevant to student's sociocultural backgrounds to maximize their academic achievement.

Multicultural education in general and multicultural math in particular, is seen as a way to improve not only education for language minority children (Ogbu, 1995), but also an imperative need for educational reform. This implies that multiple learning opportunities from all students can be provided in math classrooms. (Banks, 1994, Banks 1995) suggests a four-level approval to such reform in the education that can be applied to math and science education. This multi-level conceptualization includes that action approach, the transformation approach, the additive approach, and the contributions approach. These approaches can serve a useful tool in promoting math and science skills in multicultural settings. First, students are encouraged to take actions to solve leaning including math and science problems in terms of their meaningful way of interaction bound to the social and cultural make-up. Second, the multiculturally transformed math and science curriculum enables learners to view math and science concepts and meanings in terms of cultural diversity. Furthermore, students are afforded with added universal elements of science and math that enrich their educational opportunities in math science classrooms. Finally, students see themselves and their cultures through the contributions of pioneers and leaders in math and science who represent these students in today's classrooms.

According to current research, until recently there have not been many links to students' culture in the mathematics classroom. This may be one of the major barriers to achievement of many groups historically underrepresented in mathematics, for these students may see mathematics as a subject that has very little meaning or value for their current or future lives. Banks (1989) contends that preparing students to be functional in a competitive, pluralistic society and teaching them about their customs, heritage, history, and other aesthetic aspects are essential components of an effective educational program. As mathematics educators actively look for ways of linking mathematics and our Multicultural society (Bishop, 1988; D'Ambrosio, 1985; Frankenstein, 1990; Zaslavsky, 1993), the five dimensions of multicultural education identified by Banks (1994) provide a framework for empowering all of our students through multicultural mathematics:

1. Content integration, the illumination of key points of instruction with content reflecting diversity;
2. Knowledge construction, helping students understand how perspectives of people within a discipline influence the conclusions reached within that discipline;
3. Prejudice reduction, efforts to develop positive attitudes toward different groups;
4. Equitable pedagogy, ways to modify teaching to facilitate academic achievement among students from diverse groups; and
5. Empowering school culture and social structure, ensuring educational equality and cultural empowerment for students from diverse groups (pp. 4-5).

The inclusion of multicultural mathematics continues according to the history of research in this area simply because the growing migration, immigration and diversity of our populations demand it. However, it can be negative when it restricts ethnic groups to stereotypes and leaves us unprepared to participate in academic endeavors. It is equally negative when it waters down mathematics content in general. Multicultural education seeks to recognize the contributions, values, rights, and the equality of opportunities of all groups that compose a given society. Educators can begin to develop the ideal equality among students and build a foundation for promoting academic excellence for all students (Croom, 1997). In an earlier work, D. C. Orey stated,

> A multicultural perspective on mathematical instruction should not become another isolated topic to add to the present curriculum content base. It should be a philosophical perspective that serves as both filter and magnifier. This filter/magnifier should ensure that all students, be they from minority or majority contexts will receive the best mathematics background possible. (Orey, 1989, p.7).

We recommend that the above questions continue to be debated in order to develop inclusive paths of further development of mathematics, our societies and the schools therein. As well, we hope that the Working group may come to some consensus in regards to the confusion between what is multicultural and ethno mathematics. Sociological questions about the relationships between institutionally dominant majority and minority cultures need be reflected by our increasingly globalize world.

REFERENCES

Banks, J. (1989). *Multicultural education: Issues and perspectives.* Boston: Allyn & Bacon.

Banks, J. (1994). "Transforming the mainstream curriculum." *Educational Leadership,* 51(8), 4-8.

Banks, J. (2005). *Cultural diversity and education: Foundations, curriculum, and teaching* (5th edition). New York: Allyn & Bacon

Banks, J. (2008) *An introduction to multicultural education* (4th edition). Boston: Allyn & Bacon.

Bennett, C. (2003). *Comprehensive multicultural education: Theory and practice* (4th edition). Boston, MA: Pearson Education.

Bishop, A. (1988). "Mathematics education in its cultural context." *Educational Studies in Mathematics,* 19, 179-191.

Bishop, R. S. (1992). "Multicultural literature for children: Making informed choices." In V. Harris (Ed.), *Teaching multicultural literature in grades K-8* (pp. 37-53). Norwood, MA.

Christopher-Gordon.D'Ambrosio, U. (1985). "Ethnomathematics and Its place in the history and pedagogy of mathematics." *For the Learning of Mathematics,* 5(1), 44-48.

Croom, L. (1997). "Mathematics for all students: Access, excellence, and equity." In J. Trentacosta and M. J. Kenney (Eds.), *Multicultural and gender equity in the mathematics classrooms: The gift of diversity* (p.1-9). Reston, VA: NCTM.

Dinkins, Preston (1994) "Multicultural Teaching Strategies for simplifying mathematical concepts and principles" in C. Grant (ed) 1992 *Multicultural education for the Twenty first Century.*

Duhon-Sells, RM (1992). "The 90's perspective of multicultural education." In C. Grant (ed). *Multicultural education for the Twenty first Century* (pp. vii-x)

Frankenstein, M. (1990). "Incorporating race, class, and gender issues into a critical mathematical literacy curriculum." *The Journal of Negro Education,* 59, 336-347.

Freire, P. (1970). *Pedagogia do Oprimido* [Pedagogy of the oppressed].Rio de Janeiro: Paz e Terra.

ICME 7 – 7th International Congress in Mathematics Education (1992). Working Group 10 on Multicultural/Multilingual Classrooms. Topic subgroup 1: Curriculum, *Resources, and Materials for Multicultural/Multilingual Classrooms*, organized by Rick Patrick Scott. Quebec, Canada. In R.P. Scott (Ed.), *ICM-7 Working Group 10 on Multicultural/Multilingual Classrooms* (p.3-5). ISGEm Newsletter, Las Cruces, NM: ISGEm.

Ladson-Billings, G. (2009). *Dreamkeepers: Successful teachers of African American students* (2nd edition). San Francisco: Jossey Bass.

NAME Proceedings; The Facing the Challenge of Cultural Pluralism and Diversity: A Clarion Call for Unique Opportunities in the 21st Century Detroit, Michigan.

National Council of Teachers of Mathematics (NCTM) (1991). Professional standard for teaching mathematics. Reston, VA:Author.

Nieto, S. and P. Bode. (2009). *Affirming Diversity: The sociopolitical context of multicultural education* (5th edition). Boston: Allyn & Bacon.

Oghu, J. (1995) "Understanding cultural diversity and learning," In I. Banks and C. Banks (Eds), *Handbook of research on multucultural education*. New York. Macmillan Publishing

Orey, D.C. (1989). "Ethnomathematical perspectives on the NCTM Standards." In R.P. Scott (Ed.). *ISGEm Newsletter* (p. 5-7). Las Cruces, NM: ISGEm.

Orey, D. C. and M. Rosa (2004). *Ethnomathematics and the teaching and learning Mathematics from a multicultural perspective*. IV Festival Internacional de Matemática, San José Costa Rica 2004

Sleeter, C.E., and Grant, C.A. (1987). "An analysis of multicultural education in the United States." *Garward Education Review*, 574(4), 421-444.

Suleiman, Mahmoud F. (1997) A balance formula for math and science education in diverse settings. Presented at the NAME conference in Albuquerque NM, Oct. 29-Nov. 1997.

CHAPTER FIVE
AFTER-WORK HOURS CHALLENGES OF AMERICAN PARENTS TO ASSIST THEIR CHILDREN WITH MATH HOMEWORK AND OTHER SUBJECT MATTERS

DONNA M. GRANT

I write this chapter with the many American parents in mind, who struggle diligently and passionately during after-work hours with the challenges of assisting their children with math homework as well as other subjects. Their valiant, heartfelt and invaluable efforts are recognized, commemorated, and appreciated.

Given that learning involves a change in behavior, a pro-active student as he or she learn, will translate knowledge into action. This is the case when parents attempt to assist their children with homework, whether it is Math, Science, English, or any other subject.

Many families suffer at homework time, due to tension that arises from miscommunication between parents and students. You're shocked, you say? Okay, it's no secret that teens and parents have struggled for years with communication problems, according to Grace Fleming (Fleming, 2009). But there are ways to ease the stress when it comes to homework.

DO ANY OF THESE SITUATIONS SOUND FAMILIAR?

- The student struggles with a math problem and the parent tries to help. This only increases the stress when the student still doesn't understand.
- The student realizes at the last minute that he or she needs a poster for an assignment. The parents are tired from a long day at work, and now they have to run out for an unexpected errand. The situation is made worse because everybody will get to bed late.
- The student sits down to do homework, but tension arises when a sibling is making too much noise in the room. A fight breaks out and parents get involved.
- The student prepares to do homework and realizes he or she forgot to write down the assignment.
- The parent tells the student it's time for bed, and the student says "I haven't done my homework yet."

Many families face situations like this on a nightly basis. To ease the stress, some families can benefit by establishing a homework contract.

There are lots of payoffs to establishing a clear set of rules. Students can reap the benefits of weekly rewards, and parents can relax by avoiding the unexpected interruptions and arguments at night. Additionally, grades are likely to improve when students and parents stick to a routine. Studies have shown that homework grades improve when parents become aware of the assignments. Just by paying attention, parents improve kid's grades. Look over these suggestions for your contract.

Parents agree to:

- Provide assistance, but only when it's requested by the student.
- Provide tools and transportation necessary for homework completion.
- Reward when relevant.
- Provide a homework space.

Student agrees to:

- Maintain open communication with parents concerning assignments.
- Keep assignment logbook and don't keep any assignments a secret.
- Establish a schedule and stick to it.
- Give plenty of notice when supplies are needed.
- Bring home returned assignments and show grades to parents.

Weekly Reward Examples

- Students are free from weekend chores if all requirements are met.

- Student gets a weekly allowance.
- Saturday night free.
- Trip to the mall.
    Long Term Rewards
- A Car
- College road trip with friends
- Spring break trip
- Spending money for school clothes
- Cash for from expenses

Overall, the possibilities are endless when it comes to rewards for parents. Believe me, they will benefit just by cleaning up all the confusion and frustration at homework time. If grades improve when habits improve, parents will be overjoyed!

HOW PARENTS CAN HELP WITH HOMEWORK

According to Dr. Vincent Iannelli, research shows that parent involvement can have either a positive or negative impact on the value of homework (Iannelli, 2003). Parent involvement can be used to speed up a child's learning. Homework can involve parents in the school process. It can enhance parents' appreciation of education. It can give them an opportunity to express positive attitudes about the value of success in school.

But parent involvement may also interfere with learning. For example, parents can confuse children if the teaching techniques they use differ from those used in the classroom. Parent involvement in homework can turn into parent interference if parents complete tasks that the child is capable of completing alone.

When mothers and fathers get involved with their children's homework, communication between the school and family can improve. It can clarify for parents what is expected of students. It can give parents a firsthand idea of what students are learning and how well their child is doing in school.

Research shows that if a child is having difficulty with homework, parents should become involved by paying close attention. They should expect more requests from teachers for their help. If a child is doing well in school, parents should consider shifting their efforts to providing support for their child's own choices about how to do homework. Parents should avoid interfering in the independent completion of assignments.

Overall, homework can be an effective way for students to improve their learning and for parents to communicate their appreciation of schooling. Because a great many things influence the impact of homework achievement, expectations for homework's effects, especially in the earlier grades, must be realistic.

Homework policies and practices should give teachers and parents the flexibility to take into accounts the unique needs and circumstances of their students. That way, they can maximize the positive effects of homework and minimize the negative ones.

WHY IS IT IMPORTANT FOR MY CHILD TO LEARN MATH?

Math skills are important to a child's success – both at school and in everyday life. Understanding math also builds confidence and opens the door to a range of career options. In our everyday lives, understanding math enables us to:

- Manage time and money, and handle everyday situations that involve numbers (for example, calculate how much time we need to get to work, how much food we need in order to feed our families, predict traffic patterns to decide on the best time to travel);
- Solve problems and make sound decisions,
- Explain how we solved a problem and why we made a particular decision;
- Use technology (for example, calculators and computers) to help solve problems.

Knowing how to do math makes our day-to-day lives easier!

HOW WILL MY CHILD LEARN MATH?

Children learn math best through activities that encourage them to:
- Explore;
- think about what they are exploring;
- solve problems using information they have gathered themselves;

- explain how they reach their solutions.

Children learn easily when they can connect math concepts and procedures to their own experience. By using common household objects (such as measuring cups and spoons in the kitchen) and observing everyday events (such as weather patterns over the course of a week), they can "see" the ideas that are being taught.

An important part of learning math is learning how to solve problems. Children are encouraged to use trial and error to develop their ability to reason and to learn how to get about problem solving. They learn that there may be more than one way to solve a problem and more than one answer. They also learn to express themselves clearly as clearly as they explain their solutions.

WHAT TIPS CAN I (PARENTS) DO TO HELP MY CHILD WITH MATH?

Be Positive About Math.
- Let your child know that everyone can learn math.
- Let your child know that you think math is important and fun.
- Point out the ways in which different family members use math in their jobs.
- Be positive about your own math abilities. Try to avoid saying "I was never good at math" or "I never liked math."
- Encourage your child to be persistent if a problem seems difficult.
- Praise your child when he or she makes an effort, and share in the excitement when he or she solves a problem or understands something for the first time.

MAKE MATH PART OF YOUR CHILD'S DAY

Point out to your child the many ways in which math is used in everyday activities.
- Encourage your child to tell or show you how he or she uses math in everyday life.
- Include your child in everyday activities that involve math – making purchases, measuring ingredients, counting out plates and utensils for dinner.

- Play games and do puzzles with your child that involve math. They may focus on direction or time, logic and reasoning, sorting, or estimating.
- Do math problems with your child for fun.
- In addition to math tools, such as a ruler and a calculator, use handy household objects, such as a measuring cup and containers of various shapes and sizes, when doing math with your child.

ENCOURAGE YOUR CHILD TO GIVE EXPLANATIONS

When your child is trying to solve a problem, ask what he or she is thinking. If your child seems puzzled, ask him or her to tell you what doesn't make sense. (Talking about their ideas and how they reach solutions helps children learn to reason mathematically).

- Suggest that your child act out a problem to solve it. Have your child show how he or she reached a conclusion by drawing pictures and moving objects as well as by using words.
- Treat errors as opportunities to help your child learn something new.

TEACHING MATH TO YOUNG CHILDREN

There are a series of web pages that are available to help students understand math, and to help parents teach their children math – especially to help children have a good foundation. These web pages are not pages to teach the mere recipes or algorithms for solving problems. As an Administrator and Academician with a terminal Ph.D. Degree in Education, my philosophy is that if you understand math as you go along, you will be able to do your homework well yourself and you should be able to do well on exams. And although understanding takes some work, it is usually easier than trying to memorize rules and recall them later, especially in situations that are slightly new and different and especially under pressure of tests.

There are some things that one needs to know, almost automatically and that need to be a part of your memory, but those things can be learned with practice that should not be too tedious. Practice is important to help you do math more automatically and therefore to recognize possible solutions more quickly, but you should still understand the things you

After-Work Hours Challenges of American Parents to Assist Their Children With Math Homework and Other Subject Matters

memorize or know automatically. Memorization and rote learning, by themselves, in math just won't help much by the time you get to Algebra; you have to understand numbers and relationships between numbers – the logic of numbers and of math. It is not all that difficult if you develop a good foundation, which requires both understanding and sufficient practice to keep your understanding sharp. Contained in this chapter, I offer ways for parents or teachers to help children develop such a foundation.

The following are other essays or references that can be utilized as a point of reference on this subject for both parents and teachers. They are:

1. *"The Concept and Teaching of Place-Value"* – A theoretical explanation about (the problem of) teaching Place-Value to children, with a practical method given based on that explanation (Garlikov, 2009a).
2. *"The Socratic Method: Teaching by Asking Instead of by Telling"* – An example and explanation of how to use students' (including young children's) inherent reasoning ability to guide them gently and easily into an understanding of difficult ideas (Garlikov, 2009b).
3. *"A Supplemental Introduction to Algebra"* – Explains some things algebra books don't tend to tell students, but is important for students to know so that the can understand what algebra "is about" and see, in a sense, how it works (Garlikov, 2009c).
4. *"Understanding 'Rate" Word Problems"* – A conceptual explanation of doing various kinds of math "rate" problems (i.e., "distance/speed/time" problems, problems involving combining work done by different agents working at different rates of speed, quantity/proportion problems, etc.).

As an educator, it is my belief that there is a lack of conceptual understanding that causes students to have the most difficulty with math, along with lack of practice manipulating quantities and numbers in certain ways to the point of being comfortable and familiar with them.

For example, with regard to practice, adults often don't realize how difficult it is for young children to associate number NAMES with the QUANTITIES OF THINGS those names represent; adults have associated quantities with numbers for so long, it seems second nature to them; but to children, it is now like the following would be to an adult, and it is just as difficult:

Imagine if I said from now on we will change all the numerical names (that is 0-9) to the ten letters of the alphabet below, which you already know in order, so this should make the tasks following them even easier:

k, l, m, n, o, p, q, r, s, t.

How long do you think it would take you
- to learn to quickly recognize p objects by sight, or
- to count easily to 1 KK by m's,
- [m, o, q, s, 1K, 1m, 10, 1q, 1s, mk, mm,....tq, ts, iKK], or
- To see easily that m plus n equals p?

You would need lots of practice; and the idea would be to make the practice be enjoyable and interesting to you.

The same is true for little children with regard to learning numbers because number names don't mean anything more to children about quantities than the above letter names for quantities do to you.

I think it is important for children and their teachers to accept and believe in the following three teaching/learning principles:

1. When the student tells the parent or teacher what s/he doesn't understand or cannot do, s/he should ALSO TELL what s/he tried to do and why. S/he should tell what s/he understands about the problem and how s/he tried to go about solving it, and why s/he tried that. S/he should tell as specifically as s/he can where s/he thinks s/he is getting stuck. Obviously this will be more difficult for your children to verbalize, but they should have the opportunity and the coaxing to try to communicate in some way how they are thinking and seeing a concept or idea. It is important for parents and teachers to try to help children explain their own ideas and reasoning as much as possible.

2. It is important for the parent or teacher to (try to) understand where the child is going wrong and how the child got there, so that the teacher can correct misconceptions and so that he or she can know what sort of answer might serve the student best.

3. When a parent or teacher believes a child has understood an idea or concept, the parent or teacher should present the child with a problem that requires a slight modification of it, in order to see whether the child recognizes something new needs to be done and can see how to take it into account. If a child can do this, that presents better evidence the child does understand what s/he been doing and has not just hit on a pattern that works for particular situations or has not just learned by rote how to do what you were working with them on previously.

While teachers should be looking for any indication a student does not understand something, it is also important that the student should say if s/he does not understand a parent or teacher's explanation or answer or some part of it. There may be more the parent/teacher needs to say, or there may be some other way they need to say it so that the student can see it. "Seeing" someone else's explanation about something, particularly in math, is not always easy; and it does not mean a student cannot learn or come to understand a principle or relationship, or that a student is not smart, just because s/he doesn't see it the same way the teacher sees it, or the way the teacher says it the first time they give what they THINK is an adequate explanation. The teacher will just have to try a different approach to help, or will just have to explain more about his/her initial approach. But the student needs to let the teacher know, because otherwise the teacher may not realize s/he doesn't understand, and may keep on going in a way that gets the student really lost.

FOUNDATIONS

The following are some of the aspects of arithmetic where I believe a good foundation is particularly important. If you are the parents of young children, you can practice some of these things with your children while you are in the car with them; I found that to be a good time to work with them:

1. Naming whole numbers in order (sometimes called counting, even though one may not actually be counting anything). As young children get older, they can add more numbers to the list. Notice, naming numbers in order is different from counting, though counting sometimes involves naming numbers in order. But you can name numbers in order without counting and you can count without naming numbers in order. (You, as an adult, can count without naming numbers because you can sometimes see five objects immediately as five, without having to "count out" each one of them – as when playing dominoes; or you might multiply to get a total, or you might count by two's or by 10's, etc.) But before kids can count objects by naming the numbers one at a time in order as they point to objects, they need to learn the number names in order. So you can start with "one, two" and then add numbers as kids are able to absorb them. You can name numbers while pointing to fingers, or just reciting the numbers, or by using nursery rhymes such as "One, two, buckle my

shoe...." Give little kids plenty of practice naming numbers as high as they can go, helping them and making it fun for them, and applauding them when they learn them. (See #5 and #6 below for typical particular problems about learning umber names in order.)

2. Counting things one by one. As your children learn number names, give them practice counting things, helping them when necessary and praising them as they get it right. Counting things one by one helps them count and it reinforces the order of number names while they are young. You can have them count candy, such as M&M's, or poker chips, or the hearts (spades, clubs, diamonds) on the face cards in a deck of cards, or the dots on dice. If you have games like Chutes and Ladders or Monopoly, etc., it will give them lots of practice counting the dice and the squares they move past on the game board. Eventually the will even start to see groups of squares they won't even have to count one-by-one.

3. Naming number names by groups; e.g., by 2's, by 5's, and by 10's in particular. Once they have learned to name numbers in order, teach them to name numbers by two's, then by five's, and by tens. Once they understand WHAT it is you are teaching them, you can give them practice by the next step, #4.

4. Counting by groups; e.g., by 2's, by 5's, by 10's, etc. Make sure you get them to see how much faster it is to count out large quantities by groups, rather than one at a time. You have to point this out to most kids or they will tend to count things one at a time and not even think about counting by groups even though they know how to count by groups; they just don't think to do it, unless they have been **told and shown** at least once that **it is a faster way to count.**

5. My children had trouble learning what I would call the "transition" number names. They had trouble learning what comes after the 9's in the two digit numbers; e.g., after 29, 39, 49, etc., even though they could say numbers by tens: 10, 20, 30, 40, etc. So it took additional practice working with that in particular. I had to get them to see what came after say, 49, when they were counting by one's was the same thing that came after 40 when they were counting by 10's; that is, they needed lots of practice in seeing that when you "finished" the forties you went into the fifties, when you finished the seventies you went into the eighties, etc. So we did extra practice naming numbers starting at the 7 in each "decade"; i.e., 37, 38, 39, ? 47, 48, 49, ? 87, 88, 89, ?

6. Kids also have difficulty sometimes saying numbers in order out loud because they will accidentally jump from, something like fifty-six to seventy-seven or to sixty-seven because they get confused between changing the one's or the ten's place number. It is not a sign of any significant difficulty, but you need to watch for it so they learn not to do it.
7. Kids need to learn to read and to write numbers. This is not too difficult with single digit numbers, but it is somewhat difficult with multi-digit numbers, since the number ten, for example, written out looks like one, zero. Kids can just learn it is 10. At this stage they don't necessarily need, and might not be able to appreciate, a rationale. You can just say something like, "I know this looks like one, zero, but it is the way you write 'ten'." Similarly 11, etc. At some point, if they seem like they can follow it, you can show them that ten through nineteen all have a "1" on the left side, and that all the twenties have 2's on the left side, etc., but I wouldn't get into talking about columns or place-value. If you feel they might think it interesting, you might explain that the "teen" in each of the – teen number names is like "ten" and that the tens are like three-teen, four-teen, five-teen, and that twelve is like two-teen and so the numbers look like a ten except for the numeral that replaces the "0" in the ten. Once you get to twenty, this is easier, and you may even want to start with it – twenty one is written like twenty but with a one at the end; twenty two is like twenty with a two at the end, etc. (I will get to place-value later.)
8. When your children are very young, you can very naturally, without any fanfare, introduce them to **fractions** by breaking a cookie into roughly two parts in front of them and saying something like: "Here, I'll give you half a cookie and I will eat half [or I'll give your brother the other half]." Similarly with one-fourth of something when a reasonable occasion arises. Or you might give them "half a glass of milk" and identify it as such.
9. When your children start to study fractions in school, you can make it easier for them by explaining every fraction has two parts, which, when written, are a top number that is said first, and a bottom n umber which is said second (in the form you will have to explain to them – e.g., "four THS" instead of "four"). Let them know the bottom tells how many "pieces" you divide something up into, and the top part tells you how many of those pieces "you have" or "you are talking about." So if you divide a cookie into halves and you get one

piece, you have one half a cookie. If you have four people in your family and two of them are women, then two fourths of your family are females, You can ask them what fraction of their family they are, what fraction the children are, what fraction of the legs of a dog are front legs, or left legs, or left front legs. Etc. I find kids get a real kick out of telling you all kinds of bizarre fractions like these once they catch on to seeing how to name fractional parts of things. At some point you can also show them that fractions can be more than one whole thing, say, by breaking two cookies into halves and giving them three of the halves and asking them how many halves they have. And helping them see that three halves then is the same as one-and-a-half cookies, just as you probably already have shown them that two halves are the same as a whole cookie (except for some of the crumbs that fall when you break the cookie in half).

10. As they learn to add numbers, give them plenty of practice by letting them play games where they add numbers together. They can play with two or more dice, for example. Or they can play "double war" in cards, a game where each player turns over two cards, and the player with the highest SUM wins all the cards turned over. (When a player runs out of cards to turn over, he or she picks up the cards s/he has won and uses them. Each player keeps doing this until one player has all the cards." Or when they are old enough to start to understand the game, they can play blackjack just for fun without betting anything. They will like just trying to win each hand. As your children get better at adding and subtracting, you can show them neat "magic" tricks with numbers, such as how to add up the numbers that are on the BOTTOMS of the dice they have rolled, without having to pick up the dice to see those numbers. (The opposite sides of dice add up to 7, so if the three is rolled, a four is on the bottom; if a six is rolled, the one is on the bottom. SO if you roll two dice and get a five and a three, you know that there is a two and a four on the bottom, and can sum them up to six. Also, the opposite sides of TWO dice will add up to 14, so you could add the five and the three you see and subtract that from 14 to still get 6.) Once a kid learns how to do this trick, s/he can amaze his/her friends, and get lots of practice. Especially if using three or four dice.

11. I believe it is important that children play games that give them practice adding single digit numbers up to sums of at least 18, since 18 is the largest number you ever get when you regroup or "borrow" numbers by the "standard" subtraction "method" or recipe

After-Work Hours Challenges of American Parents to Assist Their Children With Math Homework and Other Subject Matters   57

(algorithm); e.g., if you are subtracting 9 from 38, in the standard American algorithm, you change the "thirty" to "twenty with 10 ones", and that gives you 18 ones. (If you were to get 19, you would not have had to regroup in the first lace, because you could have subtracted any digit from the 9 that you began with, without having to "borrow" to do it' e.g., if you were subtracting something from 39, you would never have to "borrow" from the "thirty", since with a 9 in the one's column, you could subtract ANY number from it in the one's column." If you are not opposed to letting them play cards, "blackjack" or 21 is an easy and excellent game for practice in developing this particular skill.

12. Children run into great difficulty learning "place-value" – what the different columns of numbers represent, AND WHY, etc. And many learn it only by rote (they never learn the "why"), which causes problems later in a number of places. I think there is an easy and great way to teach place-value, and to teach about regrouping, borrowing, etc. using poker chips with different colors. (Stacks of poker chips can also teach about fractional relationships; e.g., if you start with a stack of 32 poker chips, half of that stack is 16, half of that is 8, half of that is 4, half of that is 2, and half of that is 1; and you can show them the relationship among the stacks: e.g., 4 is half of the stack with 8 in it, and ¼ of the stack with 16. Etc., etc.) Plus, when they are first learning to count, and also learning to count by two's, etc., they can count poker chips and stack them into two's, five's, ten's. SO I recommend that you buy a pack or two of poker chips (be sure they have stacks of at least three colors – commonly red ones, white ones, and blue ones), which you can get for a few dollars a pack at some of the discount stores or at some drugstores. And I also recommend your buying two decks of cards, since you can give kids practice in counting and adding and subtracting with them. They can count cards or count the objects on the faces, or add and subtract the face values, in a number of different games they might play, or in a number of different tasks you might ask them to do, that they often will find fun.

13. If children have learned fractions and place-value, decimals will not be all that difficult, with some help and explanation. And once they have learned these things, percentages will be easy as well.

14. Finally, for now, you can lay some groundwork as early as kindergarten or first grade for work problems in general and for algebra later, by asking questions like "If I have a bag, and you have a bag, and we each have the same number of things in our bags, and

together we have four things how many things do we each have?" Let the child figure it out however s/he wants to; don't make there be some particular way to do it. As the child gets older or more sophisticated in arithmetic, you can make the question more sophisticated: "I have a bag and you have a bag, and I have twice as much as you, and together we have nine things in our bags...." Or "If we double what you have and then add three, you will have 13...." Or even harder: "I have five more than you do in my bag, but if you double what is in your bag, you will have five more than I do...." Surprisingly perhaps, kids can figure these out. Sometimes they do so by trial and error or by lucky guesses; but all of them give them more and more practice with numbers and with relationships between numbers. And they often seem to love doing these things, at least in small doses. And they also like doing "progressions", such as "if numbers start out going 1, 2, 4, 8, what number will be next, and HOW DO YOU KNOW?" You can quickly begin to make the progressions harder and they will still catch on. Or you can make two different progressions in the same problem: what should come after 3, 4, 6, 8, 12? (16) And why? (There are two progressions here: 3, 6, 12, as the first, third, and fifth numbers in the series, and 4, 8, _as the second, fourth, etc. numbers.)

15. You may have your own areas of math that you find interesting: geometry, trig, topology, etc. Try to devise games or puzzles using insights from those areas that your children might find fun to play with and think about. There are various inexpensive math puzzles, riddles, logics, or "magic" books, and free Internet sites available that teach many different aspects of math in different fun ways. Simple objects can be used to teach them math elements also. Nobel physical Richard Feynman told, for example, about how when he was still in his high chair, his father would bring home color tiles and would line them up in various ways for (and with) him so that there would be patterns, such as blue-white-blue-white-blue-white, or color patterns alternating by thirds or some other way. It was something of just a fun game for the baby, but a game that had a deeper meaning and point to his father. As long as it stays interesting or fun for the child, I do not see any harm in it, and it might have much educational developmental value for later.

WHY IS IT IMPORTANT FOR MY CHILD TO LEARN MATH?

Math skills are important to a child's success – both at school and in everyday life. Understanding math also builds confidence and opens the door to a range of career options.

In our everyday lives, understanding math enables us to:
- manage time and money, and handle everyday situations that involve numbers (for example, calculate how much time we need to get to work, how much food we need in order to feed our families, and how much money that food will cost);
- understand patterns in the world around us and make predictions based on patterns (for example, predict traffic patterns to decide on the best time to travel);
- solve problems and make sound decisions;
- explain how we solved a problem and why we made a particular decision;
- use technology (for example, calculators and computers) to help solve problems.

KNOWING HOW TO DO MATH MAKES OUR DAY-TO-DAY LIVES EASIER!

HOW WILL MY CHILD LEARN MATH?

Children learn math best through activities that encourage them to:
- explore;
- think about what they are exploring;
- solve problems using information they have gathered themselves;
- explain how they reached their solutions.

Children learn easily when they can connect math concepts and procedures to their own experience. By using common household objects (such as measuring cups and spoons in the kitchen) and observing everyday events (such as weather patterns over the course of a week), they can "see" the ideas that are being taught.

An important part of learning math is learning how to solved problems. Children are encouraged to use trial and error to develop their ability to reason and to learn how to go about problem solving. They learn that there may be more than one way to solve a problem and more than one answer. They also learn to express themselves clearly as they explain their solutions.

At school, children learn the concepts and skills identified for each grade in the Ontario mathematics curriculum in five major areas, or *strands*, or mathematics. The names of the five strands are: Number Sense and Numeration, Measurement, Geometry and Spatial Sense, Patterning and Algebra, and Data Management and Probability. You will see these strand names on your child's report card. The activities in this guide are connected with the different strands of the curriculum.

WHAT TIPS CAN I USE TO HELP MY CHILD?

BE POSITIVE ABOUT MATH!

- Let your child know that **everyone** can learn math.
- Let your child know that **you** think math is important and fun.
- Point out the ways in which different family members use math in their jobs.
- Be positive about your own math abilities. Try to avoid saying "I was never good at math" or "I never liked math."
- Encourage your child to be persistent if a problem seems difficult.
- Praise your child when he or she makes an effort, and share in excitement when he or she solves a problem or understands something for the first time.
-

MAKE MATH PART OF YOUR CHILD'S DAY

- Point out to your child that many ways in which math is used in everyday activities.
- Encourage your child to tell or show you how he or she uses math in everyday life.

- Include your child in everyday activities that involve math – making purchases, measuring ingredients, counting out plates and utensils for dinner.
- Play games and do puzzles with your child that involve math.

They may focus on direction or time, logic and reasoning, sorting, or estimating.
- Do math problems with your child for fun.
- In addition to math tools, such as a ruler and a calculator, use handy household objects, such as a measuring cup and containers of various shapes and sizes, when doing math with your child.

ENCOURAGE YOUR CHILD TO GIVE EXPLANATIONS

- When your child is trying to solve a problem, ask what he or she is thinking. If your child seems puzzled, ask him or her to tell you what doesn't make sense. (Talking about their ideas and how they reach solutions helps children learn to reason mathematically.)
- Suggest that your child act out a problem to solve it. Have your child show how he or she reached a conclusion by drawing pictures and moving objects as well as by using words.
- Treat errors as opportunities to help your child learn something new.

WHAT MATH ACTIVITIES CAN I DO WITH MY CHILD?

1. UNDERSTANDING NUMBERS

Numbers are used to describe quantities, to count, and to ad, subtract, multiply, and divide. Understanding numbers and knowing how to combine them to solve problems help us in all areas of math.
- *Count everything?* Count toys, kitchen utensils, and items of clothing as they come out of the dryer. Help your child count by pointing to and moving the objects as you say each number out loud. Count forwards and backwards from different starting places. Use household items to practice adding, subtracting, multiplying, and dividing.
- *Sing counting songs and read counting books.* Every culture has counting songs, such as "One, Two, Buckle My Shoe" and "Ten Little Monkeys", which make learning to count – both forwards and

backwards – fund for children. Counting books also capture children's imagination, by using pictures of interesting things to count and to add.

- *Discover the many ways in which numbers are used inside and outside your home.* Take your child on a "number hunt" in your home or neighbourhood. Point out how numbers are used on the television set, the microwave, and the telephone. Spot numbers in books and newspapers. Look for numbers on signs in your neighbourhood. Encourage your child to tell you whenever he or she discovers a new way in which numbers are used.
- *Ask your child to help you solve everyday number problems.* "We need six tomatoes to make our sauce for dinner, and we have only two. How many more do we need to buy?" "You have two pillows in your room and your sister has two pillows in her room. How many pillowcases do I need to wash?" "Two guests are coming to eat dinner with us. How many plates will we need?"
- *Practice "skip counting."* Together, count by 2's and 5's. Ask your child how far he or she can count by 10's. Roll two dice, one to determine a starting number and the other to determine the counting interval. Ask your child to try counting backwards from 10, 20, or even 100.
- *Make up games using dice and playing cards.* Try rolling dice and adding or multiplying the numbers that come up. Add up the totals until you reach a target number, like 100. Play the game backwards to practice subtraction.
- *Play "Broken Calculator."* Pretend that the number 8 key on the calculator is broken. Without it, how can you make the number 18 appear on the screen? (Sample answers: 20 – 2, 15 + 3). Ask other questions using different "broken" keys.

2. UNDERSTANDING MEASUREMENTS

We use measurements to determine the height, length, and width of objects, as well as the area they cover, the volume they hold, and other characteristics. We measure time and money. Developing the ability to estimate and to measure accurately takes time and practice.

- *Measure items found around the house.* Have your child find objects that are longer or shorter than a shoe or a string or ruler. Together, use a shoe to measure the length of a floor mat. Fill different containers

with sand in a sandbox or with water in the bath, and see which containers hold more and which hold less.
- *Estimate everything!* Estimate the number of steps from your front door to the edge of your yard, then walk with your child to find out how man there really are, counting steps as you go. Estimate how many bags of mild your family will need for the week. At the end of the week, count the number of bags you actually used. Estimate the time needed for a trip. If the trip is expected to take 25 minutes, when do you have to leave? Have your child count the number of stars he or she can draw in a minute. Ask if the total is more or less than your child though it would be.
- *Compare and organize household items.* Take cereal boxes or cans of vegetables from the cupboard and have your child line them up from tallest to shortest.
- *Talk about time.* Ask your child to check the time on the clock when he or she goes to school, eats meals, and goes to bed. Together, look up the time of a television program your child wants to watch. Record on a calendar the time your child's favourite away-from home activity.
- *Keep a record of the daily temperature outside and of your child's outdoor activities.* After a few weeks, ask your child to look at the record and see how the temperature affected his or her activities.
- *Include your child in activities that involve measurements.* Have your child measure the ingredients in a recipe, or the length or a bookshelf you plan to build. Trade equal amounts of money. How many pennies do you need to trade for a nickel? for a dime?

3. UNDERSTANDING GEOMETRY

- The ability to identify and describe shapes, sizes, positions, directions, and movement is important in many work situations, such as construction and design as well as in creating and understanding art. Becoming familiar with shapes and spatial relationships in their environment will help children grasp the principles of geometry in later grades.
- *Identify shapes and sizes.* When playing with your child, identify things by their shape and size: "Pass me a sugar cube." "Take the largest cereal box out of the cupboard."

- *Build structures using blocks or old boxes.* Discuss the need to build a strong base. Ask your child which shapes stack easily, and why.
- *Hide a toy and use directional language to help your child find it.* Give clues using words and phrases such as *up, down, over, under, between, through,* and *on top of.*
- *Play "I spy", looking for different shapes.* "I spy something that is round." "I spy something that is rectangular." "I spy something that looks like a cone."
- *Ask your child to draw a picture of your street, neighbourhood, or town.* Talk about where your home is in relation to a neighbour's home or the corner store. Use directional words and phrases like *beside* and *to the right of.*
- *Go on a "shape hunt."* Have your child look for as many circles, squares, triangles, and rectangles as he or she can find in the home or outside. Do the same with three dimensional objects like cubes, cones, spheres, and cylinders. Point out that street signs come in different shapes and that a pop can is like a cylinder.

4. UNDERSTANDING PATTERNS

We find patterns in nature, art, music, and literature. We also find them in numbers. Patterns are the very heart of math. The ability to recognize patterns helps us to make predictions based on our observations. Understanding patterns helps prepare children for the study of algebra in later grades.

- Look for patterns in storybooks and songs. Many children's books and songs repeat lines or passages in predictable ways, allowing children to recognize and predict the patterns.
- Create patterns using your body. Clap and stomp your foot in a particular sequence (clap, clap, stomp), have your child repeat the same sequence, then create variations of the pattern together. Teach your child simple dances that include repeated steps and movements.
- Hunt for patterns around your house and your neighbourhood. Your child will find patterns in clothing, in wallpaper, in tiles, on toys, and among trees and flowers in the park. Encourage your child to describe the patterns found. Try to identify the features of the pattern that are repeated.

- Use household items to create and extend patterns. Lay down a row of spoons pointing in different directions in a particular pattern (up, up, down, up, up, down) and ask your child to extend the pattern.
- Explore patterns created by numbers. Write the numbers from 1 to 100 in rows of 10 (1 to 10 in the first row, 11 to 20 in the second row, and son on). Note the patterns that you see when you look up and down, across, or diagonally. Pick out all the numbers that contain a 2 or a 7.

5. UNDERSTANDING AND MANAGING DATA

Every day we are presented with a vast amount of information, much of it involving numbers. Learning to collect, organize, and interpret data at an early age will help children develop the ability to manage information and make sound decisions in the future.

- *Sort household items.* As your child tidies up toys or clothing, discuss which items should go together and why. Show your child how you organize food items in the fridge – fruit together, vegetables together, drinks on one shelf, condiments on another. Encourage your child to sort other household items – crayons by color, cutlery by type or shape, coins by denomination.
- *Make a weather graph.* Have your child draw pictures on a calendar to record each day's weather. At the end of the month, make a picture graph showing how many sunny days, cloudy days, and rainy days there were in that month.
- *Make a food chart.* Create a chart to record the number of apples, oranges, bananas, and other fruit your family eats each day. At the end of the month, have your child count the number of pieces of each type of fruit eaten. Ask how many more of one kind of fruit were eaten than of another. What was your family's least favourite fruit that month?
- *Talk about the likelihood of events.* Have your child draw pictures of things your family does often, things you do sometimes, an things you never do. Discuss why you never do some things (swim outside in January). Ask your child if it's likely to rain today. It is likely that a pig will fly through the kitchen window?

WHERE CAN I GET HELP?

Many people are willing to support you in helping your child learn math, and there are also many resources available.

YOUR CHILD'S TEACHER

Your child's teacher can provide advice about helping your child with math. Here are some topics you could discuss with the teacher:
- your child's level of performance in math
- the goals your child is working towards in math, and how you can support your child in achieving them
- strategies you can use to assist your child in areas that he or she finds difficult
- activities to work on at home with our child
- Other resources, such as books, games, and websites

OTHERS WHO CAN HELP

- Consider involving relatives and friends in helping to motivate your child to learn math. Other siblings, grandparents, family friends, and your child's caregivers can add their support and encouragement.
- If your child attends a child care centre or early years centre, the staff there may be able to suggest additional math activities to do with your child.

REFERENCES

Fleming, Grace. "Homework Contract: Can Your Family Benefit?" http://ww. homeworktips. About.come/od/Studymethods/a/contract.htm. September 22, 2009.

Garlikov, Rick, "The Concept and Teaching of Place-Value", http://www.garlikov. com/Place Value.html., September 29, 2009a.

Garlikov, Rick, "The Socratic Method: Teaching by Asking Instead of by Telling", http://www.garlikov.com/soc_meth.html., September 29, 2009b.

Garlikov, Rick, "A Supplemental Introduction to Algebra," http://www.garlikov.com/ Math/Algebra.html., September 29, 2009c.

Garlikov, Rick, "Understanding 'Rate' Word Problems", http://www.garlikov.com/ math/RateTime.html., September 29, 2009.

Iannelli, Vincent, "How Parents Can Help With Homework", (I) *Pediatrics Newsletter*, Medical Review Board, September 3, 2003, p 1-2.

INDEX

academic, 1-2, 12, 15, 24, 35, 37, 39-41, 50
   achievement, 40-41
achievement, 12, 23-26, 35, 40-41, 48
activities/activity, 2-8, 12-14, 16-17, 20, 26, 48-49, 59-61, 63, 66
addition/adding, 4, 7, 11, 13, 18, 25, 30, 33, 46, 50, 54, 56-57, 61-62, 66
African American, 38, 43
arithmetic, 2, 35, 53, 58
assistance, 6, 20, 46
attitude, 3, 31, 38

career(s), 24, 48, 59
charts, 7
children, 1-3, 12-14, 16, 18-21, 25, 35, 40, 42, 45, 47-65
cognition, 15
college, 26, 33-34, 47
community, 1, 12, 15
course, 3, 23, 26-34, 49, 59
cultural, 20-21, 37-43
culture, 15, 19, 38, 40-41, 61

development, 13-14, 16-21, 37-38, 40-41, 58
   learning, 16
disadvantaged, 33-34

ego, 13
emotion(s), 6, 13-16, 18-21
ethical restraints, 13

family, 15-16, 47, 49, 56, 60, 63, 65-66
foundations, 42, 53-58
fractions, 2, 10-12, 25, 38, 55-57

geometry, 58, 60, 63
graphs, 7-9, 30
   Bar, 8
   Circle, 9
   Line, 7
   M&M®, 8

homework, 10, 28-29, 45-48, 50

id, 5, 13, 29-30, 32, 52, 55
interventions, 24

K-12, 1, 3, 12
   mathematics, 1

learning, 1-3, 12-21, 23, 25, 27, 29, 32-35, 37-40, 42-43, 45, 47-49, 51-52, 54, 57, 60-61, 65
lecture, 27, 29-31
   procedures, 29
managing data, 65

measurement(s), 9-10, 25, 60, 62-63
multicultural, 37-43
    education, 37-43
        mathematics, 38, 41
        teaching strategies, 37
multiplication, 2-4

National Department of Education, 3
neuroscience, 15, 17-18
numbers, 2, 4, 7, 10-11, 30-31, 48, 51-59, 61-62, 64-65

parent(s), 2, 12-14, 18, 20, 24, 45-53, 66
participants, 30
pattern(s), 21, 48-49, 52, 58-59, 64-65
peers, 18, 24
physical environment, 26, 32, 34
Piaget, J., 16, 18, 21
PK-12, 23, 26
problem solving, 1, 25, 35, 49, 60
procedure(s), 10, 24, 26-29, 31, 49, 59
public school(s), 3, 20

research, 2, 7, 12, 15, 17-18, 21, 23-26, 31, 33, 35, 40-41, 43, 47
rewards, 46-47

social, 1, 3, 14, 18-21, 26, 37-38, 40-41

Sousa, D., 16-17, 19, 21
statistics, 23, 26-27, 29-31, 34, 38
strategy/strategies, 2-3, 5-7, 14, 20, 24-25, 31, 34, 37, 39, 42, 66
student(s), 1-14, 18-20, 23-35, 37-48, 50-53
    African American, 38
    Black, 32
    White, 38
subtraction, 25, 56, 62
superego, 13
Sylwester, R., 19-21

teacher(s), 2-5, 7-8, 11-15, 17, 19-20, 24-25, 34-35, 37-39, 43, 47-48, 50-53, 66
teams, 4
Thompson, P., 16, 21

understanding, 2-3, 5-6, 8-9, 11, 15-16, 18, 24, 30, 39, 48, 50-51, 59, 61-65
weather, 7, 49, 59, 65
    checking for, 7, 49, 59, 65
    and managing data, 65
    measurements, 54
    numbers, 7, 59, 65
    patterns, 49, 59, 65

EDITORS

Dr. Glendolyn Duhon-Jeanlouis is currently a faculty member at Walden University and also serves as a mathematics teacher at Aldine Independent School (Texas). Duhon-Jeanlouis current research focus is on Early Childhood Education.

Dr. Alice Duhon-McCallum serves as Assistant Professor at Walden University. Her current research focus is in multi-cultural education and counseling with special interest in conflict resolution.

Dr. Ashraf Esmail is an Assistant Professor in Criminal Justice at Southern University at New Orleans. His research interests include urban/multicultural/peace education, family, cultural diversity, political sociology, criminology, social problems, and deviance.

CONTRIBUTORS

Dr. Donna Grant is currently the Interim Vice Chancellor for Student Affairs and Enrollment Services at Southern University at New Orleans. She is also an assistant professor in the College of Education. Dr. Grant retired from the public school system in Miami, Florida where she was employed as a classroom teacher, assistant principal and principal before retiring after 32 years of employment. She is the author of a children's book, *Meet Grant the Red Bear*, and a phonics program, Phonics Made Simple (PMS). Dr. Grant is a member of Delta Sigma Theta Sorority and Sweet Home Missionary Baptist Church in Miami, Florida.

Dr. Doris L. Terrell is an associate professor and Chair of the Curriculum of the School of Education at Clark Atlanta University. Her research interests include curriculum and instruction, peace/ant-violence education, wellness, and multicultural education. She has served as Past President for the National Association for Peace/Anti-Violence Education.

Dr. Gwendolyn Duhon serves as Associate Professor of Education at McNeese State University. She is published in the areas of self-efficacy and teacher preparation.

Dr. Jas Sullivan is an Assistant Professor of Political Science and African & American Studies in the Department of Political Science at Louisiana State University. His research interests are black politics, political psychology, racial identity, and stereotypes.

Dr. Lisa Eargle is an Associate Professor of Sociology as Francis Marion University. She teaches a wide variety of courses in Stratification and Inequality, Population Dynamics, and Research Methods. Her research interests include community development and adaptation, social justice issues, and educational processes.

Dr. Mattie Spears is a Professor of Education with Duplichain University, an online distance learning institution that provides Masters and PhD's degree in Criminal Justice and Education. She earned the Ed.D.

in Organizational Leadership from Nova Southeastern University in 2006. She completed her Bachelor's degree in Education at Grambling State University in Grambling, Louisiana and Master's in Counseling at Southern University in Baton Rouge, Louisiana. Her research interests include Effective Classroom Management, Curriculum and Instruction, Educational Leadership, and Differentiated Instruction.

Dr. Melba Venison is Dean at Duplichain University, an online distance learning institution that provides Masters and PhD's degree in Criminal Justice and Education. Her research interests are focused on areas in reading.

Dr. Rose Duhon-Sells received her PhD in Curriculum and Instruction from Kansas State University. She is Vice-Chancellor of Academic Affairs at Duplichain University. Dr. Duhon-Sells has over 20 books and 50 articles to her credit. She has presented her research across the world, and is the founder of the National Association for Multi-Cultural Education (NAME) and the National Association for Peace/Anti-Violence Education (NAPE).

www.ingramcontent.com/pod-product-compliance
Lightning Source LLC
Chambersburg PA
CBHW030118010526
44116CB00005B/311